Complex Numbers
and
Geometry

Liang-shin Hahn

SPECTRUM SERIES

The Spectrum Series of the Mathematical Association of America was so named to reflect its purpose: to publish a broad range of books including biographies, accessible expositions of old or new mathematical ideas, reprints and revisions of excellent out-of-print books, popular works, and other monographs of high interest that will appeal to a broad range of readers, including students and teachers of mathematics, mathematical amateurs, and researchers.

Mathematical Association of America
1529 Eighteenth Street, NW
Washington, DC 20036
800-331-1MAA   FAX 202-265-2384

Complex Numbers
and
Geometry

Liang-shin Hahn

Published by
The Mathematical Association of America

LaTeX macros by Michael Downes

©1994 by
The Mathematical Association of America (Incorporated)
Library of Congress Catalog Card Number 93-79038

ISBN 0-88385-510-0

Printed in the United States of America

Current Printing (last digit):
10 9 8 7 6 5 4 3 2 1

To my parents

Shyr-Chyuan Hahn, M.D., Ph.D.
Shiu-Luan Tsung Hahn

And to my wife

Hwei-Shien Lee Hahn, M.D.

Preface

The shortest path between two truths in the real domain passes through the complex domain. — J. Hadamard

This book is the outcome of lectures that I gave to prospective high-school teachers at the University of New Mexico during the Spring semester of 1991. I believe that while the axiomatic approach is very important, too much emphasis on it in a beginning course in geometry turns off students' interest in this subject, and the chance for them to appreciate the beauty and excitement of geometry may be forever lost. In our high schools the complex numbers are introduced in order to solve quadratic equations, and then no more is said about them. Students are left with the impression that complex numbers are artificial and not really useful and that they were invented for the sole purpose of being able to claim that we can solve every quadratic equation. In reality, the study of complex numbers is an ideal subject for prospective high-school teachers or students to pursue in depth. The study of complex numbers gives students a chance to review number systems, vectors, trigonometry, geometry, and many other topics that are discussed in high school, not to mention an introduction to a unified view of elementary functions that one encounters in calculus.

Unfortunately, complex numbers and geometry are almost totally neglected in our high-school mathematics curriculum. The purpose of the book is to demonstrate that these two subjects can be blended together beautifully, resulting in easy proofs and natural generalizations of many theorems in plane geometry—such as the Napoleon theorem, the Simson theorem, and the Morley theorem. In fact, one of my students told me that she can not imagine that anyone who fails to become excited about the material in this book could ever become interested in mathematics.

The book is self-contained—no background in complex numbers is assumed—and can be covered at a leisurely pace in a one-semester course. Chapters 2 and 3 can be read independently. There are over 100 exercises, ranging from muscle exercises to brain exercises and readers are strongly urged to try at least half of these exercises. All the elementary geometry one needs to read this book can be found in Appendix A. The most sophisticated tools used in the book are the addition formulas for the sine and cosine functions and determinants of order 3. On several occasions matrices are mentioned, but these are supplementary in nature and those readers who are unfamiliar with matrices may safely skip these paragraphs. It is my belief that the book can be used profitably by high-school students as enrichment reading.

It is my pleasure to express heartfelt appreciation to my colleagues and friends, Professors Jeff Davis, Bernard Epstein, Reuben Hersh, Frank Kelly, and Ms. Moira Robertson, all of whom helped me with my awkward English on numerous occasions. (English is not my mother tongue.) Also, I want to express gratitude to my three sons, Shin-Yi, Shin-Jen and Shin-Hong, who read the entire manuscript in spite of their own very heavy schedules, corrected my English grammar, and made comments from quite different perspectives, which resulted in considerable improvement. Furthermore, I want to thank Ms. Linda Cicarella and Ms. Gloria Lopez, who helped me with LaTeX, which is used to type the manuscript. Linda also prepared the index of the book. Last but not least, I am deeply grateful to Professor Roger Horn, the chair of the Spectrum Editorial Board, for his patience in correcting my English, and for his very efficient handling of my manuscript.

L.-s. H.

Contents

CHAPTER I

Complex Numbers

I.I Introduction to Imaginary Numbers

One of the most important properties of the real numbers is that the operations of addition, subtraction, multiplication and division can be carried out freely (with the exception of division by 0). Because of this, an arbitrary linear equation

$$ax + b = 0 \qquad (a \neq 0)$$

can be solved within the realm of real numbers as $x = -b/a$. However, the situation is quite different for quadratic equations. For example, a quadratic equation

$$x^2 + 1 = 0$$

cannot be solved for x within the realm of real numbers. The square of any real number cannot be negative, so

$$x^2 + 1 \geq 1 > 0 \quad \text{for any real number } x.$$

Therefore, $x^2 + 1 = 0$ is impossible for any real number x. In a situation like this, we extend the realm of the number system so that the equation becomes solvable. For instance, for a kindergarten child who knows only positive integers, an equation such as

$$7 + \boxed{} = 3$$

is unreasonable, and for persons who know only integers, $5x = 2$ and $x^2 = 17$ have no solution. But by extending our number system to include negative numbers, fractions, and irrational numbers, these equations have solutions -4, $2/5$, $\pm\sqrt{17}$, respectively.

The situation is pretty much the same for $x^2 + 1 = 0$. We extend our number system to include numbers such as $\sqrt{-1}$; i.e., a number whose square is -1. Such numbers are not quite in agreement with our intuition, and many mathematicians in the past objected to introducing such monsters, so they are called *imaginary numbers*. It was not until the 18th century, with skillful manipulations of imaginary numbers, that L. Euler (1707–1783) obtained numerous interesting results. By representing the imaginary numbers as points in a plane, C. F. Gauss (1777–1855) renamed them as *complex numbers*, and applied them to obtain many remarkable results in number theory, thus establishing the citizenship of complex numbers in the number system. About the same time, in trying to find a uniform way of computing definite integrals, A. L. Cauchy (1789–1857) investigated the differential and integral calculus of functions with complex numbers as variables. This is the genesis of the theory of functions that cultivated the environment for N. H. Abel (1802–1829) and C. G. J. Jacobi (1804–1851) to discover the elliptic functions. Furthermore, the development of projective geometry shows that the complex numbers are indispensable in geometry as well. As research progresses, it is now clear that to truly understand mathematics, even merely calculus, the realm of the real numbers is unnaturally narrow, and it is imperative that we work with complex numbers to attain uniformity and harmonicity.

It is customary to use the first letter i of the word 'imaginary' for $\sqrt{-1}$. Thus complex numbers are numbers of the form $a + ib$, where a and b are real numbers, and computations with them are carried out just as with real numbers if we remember to replace i^2 by -1. For example,

$$(a + ib) \pm (c + id) = (a \pm c) + i(b \pm d),$$

$$(a + ib) \cdot (c + id) = ac + ibc + iad + i^2bd$$

$$= (ac - bd) + i(bc + ad).$$

The division of two complex numbers, $(a + ib)/(c + id)$, involves finding a complex number $x + iy$ satisfying

$$a + ib = (c + id) \cdot (x + iy).$$

Hence, by the above computation, we get

$$a + ib = (cx - dy) + i(dx + cy);$$

and so it is sufficient to find x and y satisfying

$$cx - dy = a, \qquad dx + cy = b.$$

This system of simultaneous equations has a unique solution

$$x = \frac{ac + bd}{c^2 + d^2}, \qquad y = \frac{bc - ad}{c^2 + d^2},$$

unless $c = d = 0$. Hence

$$\frac{a + ib}{c + id} = \frac{ac + bd}{c^2 + d^2} + i\frac{bc - ad}{c^2 + d^2}.$$

Of course, this can be obtained by multiplying both the numerator and the denominator by $c - id$.

But why are such operations justified? Isn't the addition of a real number a and an imaginary number ib to get $a + ib$ similar to the addition of $17m^2$ and $4kg$ to get 21^0C? Also, $x^2 + 1 = 0$ has two solutions, and which one of them is i? Note that $x^2 - 1 = 0$ also has two solutions, the positive one is 1 and the other -1. But is it meaningful to say i is positive?

1.2 Definition of Complex Numbers

To answer the criticism at the end of the last section, we now give a formal definition of complex numbers. But first let us recall the properties of the real number system $\mathbb{R}$.

(I) Properties concerning addition.

Two arbitrary real numbers a and b uniquely determine a third number called their *sum*, denoted by $a + b$, with the following properties:

A_1. *Commutative law*: $a + b = b + a$ for all $a, b \in \mathbb{R}$.

A_2. *Associative law*: $(a + b) + c = a + (b + c)$ for all $a, b, c \in \mathbb{R}$.

A_3. *Additive identity*: There is a unique real number, denoted 0, such that

$$a + 0 = 0 + a = a \quad \text{for all } a \in \mathbb{R}.$$

A_4. *Additive inverse*: For every $a \in \mathbb{R}$, there is a unique $x \in \mathbb{R}$ satisfying

$$a + x = x + a = 0.$$

This unique solution will be denoted by $-a$.

(II) Properties concerning multiplication.

Two arbitrary real numbers a and b uniquely determine a third number called their *product*, denoted by ab, with the following properties:

M_1. *Commutative law*: $ab = ba$ for all $a, b \in \mathbb{R}$.

M_2. *Associative law*: $(ab)c = a(bc)$ for all $a, b, c \in \mathbb{R}$.

M_3. *Multiplicative identity*: There is a unique real number, denoted 1, such that

$$a \cdot 1 = 1 \cdot a = a \quad \text{for all } a \in \mathbb{R}.$$

M_4. *Multiplicative inverse*: For every $a \in \mathbb{R}$, $a \neq 0$, there is a unique number $x \in \mathbb{R}$ satisfying

$$ax = xa = 1.$$

This unique solution will be denoted by $\frac{1}{a}$ or a^{-1}.

(III) Distributive law.

$$a(b + c) = ab + ac \quad \text{for all} \quad a, b, c \in \mathbb{R}.$$

Any set that satisfies these properties is called a *field*. Thus the set $\mathbb{R}$ of all real numbers is a field. Similarly, the set $\mathbb{Q}$ of all rational numbers forms a field. However, neither the set $\mathbb{Z}$ of all integers, nor the set $\mathbb{N}$ of all natural numbers is a field.

In the previous section, we said complex numbers are numbers of the form $a + ib$, where a and b are real numbers. Thus complex numbers are essentially a pair of real numbers a and b. Therefore, we give a formal definition as follows.

DEFINITION 1.2.1. *A **complex number** is an ordered pair (a, b) of real numbers with the following properties: Two complex numbers (a, b) and (c, d) are equal if and only if $a = c$ and $b = d$. The sum and product of two complex numbers (a, b) and (c, d) are defined by*

$$(a, b) + (c, d) = (a + c, \ b + d),$$

$$(a, b) \cdot (c, d) = (ac - bd, \ bc + ad).$$

Note that our definition of equality for complex numbers has the following properties:

(a) *Reflexive:* $(a, b) = (a, b)$ for every complex number (a, b);

(b) *Symmetric:* $(a, b) = (c, d) \iff (c, d) = (a, b)$;

(c) *Transitive:* $(a, b) = (c, d), \quad (c, d) = (e, f) \implies (a, b) = (e, f)$.

THEOREM 1.2.2. *With addition and multiplication defined as above, the set $\mathbb{C}$ of all complex numbers is a field.*

Proof. A muscle exercise. $\qquad \square$

Now, if we consider complex numbers of the form $(a, 0)$, then

$$(a, 0) \pm (b, 0) = (a \pm b, 0);$$

$$(a, 0) \cdot (b, 0) = (ab, 0);$$

$$\frac{(a,0)}{(b,0)} = \left(\frac{a}{b},0\right) \qquad \text{(provided} \quad b \neq 0\text{),}$$

which is identical to the operations between two real numbers a and b. In other words, there will be no confusion if we regard a complex number of the form $(a,0)$ as a real number a. Consequently, we shall consider real numbers to be particular complex numbers whose second component is zero.

Next, consider the complex number $(0,1)$. We have

$$(0,1)^2 = (0,1) \cdot (0,1) = (-1,0) = -1.$$

Namely, the complex number $(0,1)$ corresponds to $\sqrt{-1}$ in the previous section. Naturally, the square of $(0,-1)$ is also -1, but if we denote $(0,1) = i$, then an arbitrary complex number (a,b) can be rewritten as

$$(a,b) = (a,0) + (0,b) = (a,0) + (0,1) \cdot (b,0),$$

which justifies the expression $a + ib$.

The complex number i is called the *imaginary unit*. For a complex number $\alpha = (a,b) = a + ib$ $(a,b \in \mathbb{R})$, a is called the *real part* of the complex number α and is denoted by $\Re\alpha$; similarly, b is called the *imaginary part* of α, and is denoted by $\Im\alpha$. Thus real numbers are complex numbers whose imaginary parts are 0. On the other hand, complex numbers whose real parts are 0 are called *purely imaginary*. Note carefully, both the real part *and* the imaginary part of a complex number are real numbers!

For a complex number $\alpha = (a,b) = a + ib$, the number $(a,-b) = a - ib$ is called the *complex conjugate* or the *conjugate complex number* of α, and is denoted by $\overline{\alpha}$. The following relations are easy to verify:

$$\overline{\alpha \pm \beta} = \overline{\alpha} \pm \overline{\beta};$$

$$\overline{\alpha\beta} = \overline{\alpha} \cdot \overline{\beta};$$

$$\overline{\left(\frac{\alpha}{\beta}\right)} = \frac{\overline{\alpha}}{\overline{\beta}} \qquad (\beta \neq 0);$$

$$\Re\alpha = \frac{\alpha + \overline{\alpha}}{2}; \quad \Im\alpha = \frac{\alpha - \overline{\alpha}}{2i};$$

$$\overline{\overline{\alpha}} = \alpha.$$

For any complex number $\alpha = a + ib \, (a, b \in \mathbb{R})$, the product

$$\alpha\overline{\alpha} = a^2 + b^2$$

is always real and nonnegative. Its nonnegative square root is called the *modulus* or the *absolute value* of the complex number α, and is denoted by $|\alpha|$. Thus

$$|\alpha| = \sqrt{a^2 + b^2} = (\alpha\overline{\alpha})^{1/2}.$$

THEOREM 1.2.3. $|\alpha| = 0$ *if and only if* $\alpha = 0$.

Proof. Let $\alpha = a + ib \, (a, b \in \mathbb{R})$. Then $|\alpha|^2 = a^2 + b^2$. Therefore,

$$|\alpha| = 0 \Longleftrightarrow a^2 + b^2 = 0.$$

But $a^2 \geq 0$, $b^2 \geq 0$ for any real numbers a and b, hence

$$a^2 + b^2 = 0 \Longleftrightarrow a^2 = 0, \quad b^2 = 0$$

$$\Longleftrightarrow a = 0, \quad b = 0$$

$$\Longleftrightarrow \alpha = 0.$$

$\square$

Note that we have used the fact that a and b are real numbers, otherwise $a^2 + b^2 = 0$ would not imply $a = b = 0$. For example, let $a = 1, b = i$, then $a^2 + b^2 = 0$, but neither $a = 0$ nor $b = 0$.

It is simple to verify that

$$|\Re\alpha| \leq |\alpha|, \quad |\Im\alpha| \leq |\alpha|, \quad |\overline{\alpha}| = |\alpha|;$$

$$|\alpha\beta| = |\alpha| \cdot |\beta| \quad \text{(in particular, } |-\alpha| = |\alpha|\text{)};$$

$$\left|\frac{\alpha}{\beta}\right| = \frac{|\alpha|}{|\beta|} \quad \text{(provided } \beta \neq 0\text{)}.$$

THEOREM 1.2.4. *For any complex numbers α and β,*

$$\alpha\beta = 0 \Longleftrightarrow \alpha = 0 \quad or \quad \beta = 0.$$

Equivalently,

$$\alpha\beta \neq 0 \Longleftrightarrow \alpha \neq 0 \quad and \quad \beta \neq 0.$$

Proof. By the previous theorem,

$$\alpha\beta = 0 \Longleftrightarrow |\alpha\beta| = 0$$

$$\Longleftrightarrow |\alpha| \cdot |\beta| = 0.$$

Since $|\alpha|$ and $|\beta|$ are real numbers,

$$|\alpha| \cdot |\beta| = 0 \Longleftrightarrow |\alpha| = 0 \quad or \quad |\beta| = 0$$

$$\Longleftrightarrow \alpha = 0 \quad or \quad \beta = 0.$$

$\square$

Remark. In a set where multiplication is defined, if $\alpha\beta = 0$ even though $\alpha \neq 0$, $\beta \neq 0$, then α and β are called *zero divisors*. The previous theorem asserts that *the complex field $\mathbb{C}$ does not have a zero divisor.*

There are algebraic systems that have zero divisors. For example, consider the set of all 2×2 matrices:

$$\left\{ \begin{pmatrix} a & b \\ c & d \end{pmatrix}; \quad a, b, c, d \in \mathbb{R} \right\}.$$

Its addition and multiplication are defined as

$$\begin{pmatrix} a & b \\ c & d \end{pmatrix} + \begin{pmatrix} a' & b' \\ c' & d' \end{pmatrix} = \begin{pmatrix} a + a' & b + b' \\ c + c' & d + d' \end{pmatrix};$$

$$\begin{pmatrix} a & b \\ c & d \end{pmatrix} \cdot \begin{pmatrix} a' & b' \\ c' & d' \end{pmatrix} = \begin{pmatrix} aa' + bc' & ab' + bd' \\ ca' + dc' & cb' + dd' \end{pmatrix}.$$

The zero element is

$$0 = \begin{pmatrix} 0 & 0 \\ 0 & 0 \end{pmatrix}.$$

Then, even though

$$\begin{pmatrix} 0 & 1 \\ 0 & 2 \end{pmatrix} \neq 0 \quad \text{and} \quad \begin{pmatrix} 3 & 4 \\ 0 & 0 \end{pmatrix} \neq 0,$$

we have

$$\begin{pmatrix} 0 & 1 \\ 0 & 2 \end{pmatrix} \cdot \begin{pmatrix} 3 & 4 \\ 0 & 0 \end{pmatrix} = 0.$$

Note that in the proof of the theorem, we have used the fact that *the real field* $\mathbb{R}$ *has no zero divisor.*

1.3 Quadratic Equations

As we saw in §1.2, the complex number $i = (0, 1)$ satisfies the equation $x^2 + 1 = 0$. But how about other quadratic equations? Do they have solutions in $\mathbb{C}$? Or, do we have to keep on adding new members to our number system? Consider the following.

EXAMPLE. Let us find the roots of

$$\frac{1}{2}x^2 + (1 + i)x - i = 0.$$

Solution. By completing the square, we get

$$\{x + (1 + i)\}^2 = 2i + (1 + i)^2 = 4i.$$

$$\therefore \ x + (1 + i) = \pm 2\sqrt{i}.$$

$$x = -(1 + i) \pm 2\sqrt{i}.$$

But what is the square root of the imaginary unit i? It must be a number whose square is i. So set

$$u + iv = \sqrt{i} \qquad (u, v \in \mathbb{R}).$$

Then, squaring both sides, we get

$$(u^2 - v^2) + 2uvi = i.$$

$$\therefore \ u^2 - v^2 = 0, \qquad uv = \frac{1}{2}.$$

From the first equation, we get $u = \pm v$. Suppose $u = v$, then from the second equation, we get

$$u = v = \pm\frac{\sqrt{2}}{2};$$

and so

$$\sqrt{i} = u + iv = \pm\frac{\sqrt{2}}{2}(1 + i).$$

The case $u = -v$ would not happen since it would imply $u^2 = -\frac{1}{2}$, but this is impossible for u real.[1] It follows that

$$x = -(1 + i) \pm \sqrt{2}(1 + i) = (-1 \pm \sqrt{2})(1 + i).$$

So, it was not necessary to extend our complex number system in order to solve this quadratic equation. (The reader should verify that the results obtained actually satisfy the quadratic equation.)

Let us now consider a general quadratic equation

$$ax^2 + bx + c = 0 \qquad (a \neq 0).$$

---

[1] Neglecting the condition that u is real, if we proceed to solve $u^2 = -\frac{1}{2}$, we get

$$u = \pm\frac{\sqrt{2}}{2}i, \qquad v = \mp\frac{\sqrt{2}}{2}i.$$

Hence,

$$\sqrt{i} = u + iv = \pm\frac{\sqrt{2}}{2}i(1 - i) = \pm\frac{\sqrt{2}}{2}(1 + i),$$

so we obtain the same result as before.

Since we have extended our number system to complex numbers, we should discuss the case $a, b, c \in \mathbb{C}$. Since complex numbers obey the same rules as real numbers as far as addition, subtraction, multiplication and division are concerned, by dividing both sides of the above equation by a and completing the square as in the case of real coefficients, we get

$$\left(x + \frac{b}{2a}\right)^2 = \frac{b^2 - 4ac}{4a^2}.$$

Setting

$$z = x + \frac{b}{2a}, \qquad \zeta = \frac{b^2 - 4ac}{4a^2},$$

our problem becomes whether $z^2 = \zeta$ can be solved for an arbitrary complex number ζ. Let

$$z = u + iv, \qquad \zeta = \alpha + i\beta.$$

Our problem can be restated as: Can we always find a pair of real numbers u and v such that

$$(u + iv)^2 = \alpha + i\beta$$

for an arbitrary pair of real numbers α and β? Rewriting the last equality, we get

$$(u^2 - v^2) + 2iuv = \alpha + i\beta.$$

Hence our problem reduces to solving the system of simultaneous equations

$$u^2 - v^2 = \alpha, \qquad 2uv = \beta.$$

Since

$$\left(u^2 + v^2\right)^2 = \left(u^2 - v^2\right)^2 + \left(2uv\right)^2 = \alpha^2 + \beta^2,$$

and $u^2 + v^2 \geq 0$, $\alpha^2 + \beta^2 \geq 0$ $(\because u, v, \alpha, \beta \in \mathbb{R})$, we get

$$u^2 + v^2 = \sqrt{\alpha^2 + \beta^2}.$$

It follows that

$$u^2 = \frac{1}{2}\left(\alpha + \sqrt{\alpha^2 + \beta^2}\right), \qquad v^2 = \frac{1}{2}\left(-\alpha + \sqrt{\alpha^2 + \beta^2}\right).$$

Note that

$$\frac{1}{2}\left(\alpha + \sqrt{\alpha^2 + \beta^2}\right) \geq 0, \qquad \frac{1}{2}\left(-\alpha + \sqrt{\alpha^2 + \beta^2}\right) \geq 0,$$

and so

$$u = \pm\left(\frac{\alpha + \sqrt{\alpha^2 + \beta^2}}{2}\right)^{1/2}, \qquad v = \pm\left(\frac{-\alpha + \sqrt{\alpha^2 + \beta^2}}{2}\right)^{1/2},$$

where the signs must be chosen to satisfy $2uv = \beta$. That is, the square root $\sqrt{\zeta} = u + iv$ is given by

$$\sqrt{\zeta} = \begin{cases} \pm\left(\left(\frac{\alpha+\sqrt{\alpha^2+\beta^2}}{2}\right)^{1/2} + i\left(\frac{-\alpha+\sqrt{\alpha^2+\beta^2}}{2}\right)^{1/2}\right), & \text{for } \beta > 0; \\[2ex] \pm\left(-\left(\frac{\alpha+\sqrt{\alpha^2+\beta^2}}{2}\right)^{1/2} + i\left(\frac{-\alpha+\sqrt{\alpha^2+\beta^2}}{2}\right)^{1/2}\right), & \text{for } \beta < 0; \\[2ex] \pm\sqrt{\alpha}, & \text{for } \beta = 0,\ \alpha \geq 0; \\[1ex] \pm i\sqrt{-\alpha}, & \text{for } \beta = 0,\ \alpha < 0. \end{cases}$$

We have shown that *every (nonzero) complex number has two square roots.*

Remark. For $\xi \in \mathbb{R}$, the notation $\sqrt{\xi}$ was used for the nonnegative square root when $\xi \geq 0$, and $\sqrt{\xi} = i\sqrt{-\xi}\ (= i\sqrt{|\xi|})$ when $\xi < 0$. However, in the sequel, for a nonreal complex number ξ, the notation $\sqrt{\xi}$ shall simply mean a square root of ξ and shall not stand for one particular square root.

By our above convention, when ξ and η are negative real numbers, the equality

$$\sqrt{\xi} \cdot \sqrt{\eta} = \sqrt{\xi \cdot \eta}$$

is no longer valid. It is valid if we interpret both sides as *sets* of complex numbers.

We have established the following

THEOREM 1.3.1. *A quadratic equation*

$$ax^2 + bx + c = 0, \qquad a, b, c \in \mathbb{C}, \quad a \neq 0,$$

has two roots in $\mathbb{C}$, *which are given by*

$$\frac{-b \pm \sqrt{b^2 - 4ac}}{2a}.$$

1.4 Significance of the Complex Numbers

In the previous section, we have seen that every quadratic equation has solutions in the complex field $\mathbb{C}$. But how about cubic equations, quartic equations, etc.? Do we have to expand our number system each time we deal with higher degree equations? One of the beauties of the complex number system is the validity of the following

THEOREM 1.4.1 (Fundamental Theorem of Algebra). *A polynomial equation*

$$a_0 x^n + a_1 x^{n-1} + \cdots + a_{n-1} x + a_n = 0,$$

where $a_k \in \mathbb{C}$ $(k = 0, 1, 2, \ldots, n)$, $a_0 \neq 0$, $n \geq 1$, *has a solution in* $\mathbb{C}$. *In other words,*

$$\mathbb{C} \text{ is algebraically closed.}$$

The above equation is called a polynomial equation of degree n (when $a_0 \neq 0$). From the fundamental theorem of algebra, we have the following

COROLLARY 1.4.2. *A polynomial equation of degree* n *has exactly* n *roots in* $\mathbb{C}$ *taking the multiplicities into account.*

EXAMPLE. Solve the cubic equation $z^3 + i = 0$.

Solution. Let $z = u + iv$ $(u, v \in \mathbb{R})$. Then, since

$$(u + iv)^3 = u^3 + 3iu^2 v - 3uv^2 - iv^3,$$

we obtain

$$\begin{cases} u^3 - 3uv^2 = 0; \\ 3u^2v - v^3 + 1 = 0. \end{cases}$$

From the first of these equations, we get

$$u(u^2 - 3v^2) = 0.$$

Hence

$$u = 0 \quad \text{or} \quad u^2 - 3v^2 = 0.$$

When $u = 0$, from the second equation, we get

$$v^3 - 1 = 0, \qquad (v - 1)(v^2 + v + 1) = 0.$$

Since $v \in \mathbb{R}$, $v^2 + v + 1 \neq 0$, and so $v = 1, \therefore z = i$. When $u^2 - 3v^2 = 0$, $u = \pm\sqrt{3}v$. Substituting this into the second equation, we get

$$3\left(\pm\sqrt{3}v\right)^2 v - v^3 + 1 = 0, \quad \text{i.e.,} \quad 8v^3 + 1 = 0.$$

$$\therefore (2v + 1)(4v^2 - 2v + 1) = 0.$$

Since $v \in \mathbb{R}$, $4v^2 - 2v + 1 \neq 0$, and so $v = -\frac{1}{2}$. Hence

$$u = \mp\frac{\sqrt{3}}{2}, \qquad z = \frac{\pm\sqrt{3} - i}{2}.$$

Therefore, we get 3 solutions

$$z = i, \quad \frac{\pm\sqrt{3} - i}{2}.$$

C. F. Gauss (1777–1855) gave several proofs of the fundamental theorem of algebra in his dissertation. Readers who are interested in its proof may consult standard textbooks in complex analysis such as J. Bak and D. J. Newman's *Complex Analysis* [Springer-Verlag, New York, 1982], or R. P. Boas's *Invitation to Complex Analysis* [Random House, New York, 1987]. Note that the fundamental theorem of algebra asserts the existence of solutions in $\mathbb{C}$, but does not tell us how to find them. In fact, there is no *algebraic* formula that works for *every* quintic polynomial (or higher).

1.5 Order Relation in the Complex Field

We can always compare the magnitudes of any two real numbers. That is, given $a, b \in \mathbb{R}$, either $a > b$, or $a = b$, or $b > a$. Can this result be extended to complex numbers? To answer such a question, we first reexamine the order relation in $\mathbb{R}$.

P_1. (Trichotomy) For any $a \in \mathbb{R}$, one and only one of the following three relations holds:

$$a > 0, \qquad a = 0, \qquad -a > 0.$$

For $a, b \in \mathbb{R}$, if we define

$$a > b \qquad \text{if and only if} \quad a - b > 0,$$

then P_1 is equivalent to the assertion that for any $a, b \in \mathbb{R}$, one and only one of the following three relations holds:

$$a > b, \qquad a = b, \qquad b > a.$$

Furthermore, the order relation in $\mathbb{R}$ satisfies the following:

P_2. $a > 0, \ b > 0 \Longrightarrow a + b > 0$.

P_3. $a > 0, \ b > 0 \Longrightarrow ab > 0$.

It turns out that all the properties of the order relation in $\mathbb{R}$, such as

$$a > b, \ b > c \Longrightarrow a > c;$$

$$a > b, \ c > 0 \Longrightarrow ac > bc;$$

$$a > b, \ c < 0 \Longrightarrow ac < bc$$

follow from the positivity postulates P_1, P_2, and P_3. In other words, an order relation is useful only if all three postulates P_1, P_2, P_3 are satisfied.

THEOREM 1.5.1. *It is possible to extend the order relation in $\mathbb{R}$ to $\mathbb{C}$ such that P_1 and P_2 are satisfied, but it is impossible to satisfy P_3.*

Proof. For $\alpha = a + ib$ $(a, b \in \mathbb{R})$, define

$$\alpha > 0 \iff \begin{cases} a > 0, & \text{or} \\ a = 0 & \text{and} \quad b > 0. \end{cases}$$

P_1. For any $\alpha = a + ib$ $(a, b \in \mathbb{R})$, one of the following must hold:

$$a > 0, \qquad a = 0, \qquad -a > 0.$$

(a) If $a > 0 \implies \alpha > 0$.

(b) If $-a > 0 \implies -\alpha > 0$.

(c) If $a = 0$, $\begin{cases} b > 0 \implies \alpha > 0; \\ b = 0 \implies \alpha = 0; \\ -b > 0 \implies -\alpha > 0. \end{cases}$

We have shown that one and only one of

$$\alpha > 0, \qquad \alpha = 0, \qquad -\alpha > 0$$

holds for an arbitrary $\alpha \in \mathbb{C}$.

P_2. Suppose $\alpha > 0$, $\alpha' > 0$, where

$$\alpha = a + ib, \qquad \alpha' = a' + ib' \qquad (a, b, a', b' \in \mathbb{R}).$$

Then

$$a > 0 \quad \text{or} \quad \{a = 0 \text{ and } b > 0\},$$

$$a' > 0 \quad \text{or} \quad \{a' = 0 \text{ and } b' > 0\}.$$

We must check all combinations of these cases:

(a) $a > 0$, $a' > 0 \implies a + a' > 0 \implies \alpha + \alpha' > 0$.

(b) $a > 0$, $\{a' = 0 \text{ and } b' > 0\} \implies a + a' > 0 \implies \alpha + \alpha' > 0$.

(c) $a' > 0$, $\{a = 0 \text{ and } b > 0\} \implies a + a' > 0 \implies \alpha + \alpha' > 0$.

(d) Finally, $\{a = 0 \text{ and } b > 0\}$ and $\{a' = 0 \text{ and } b' > 0\}$, then $a + a' = 0$ and $b + b' > 0$, and so $\alpha + \alpha' > 0$.

P_3. Suppose the order relation in $\mathbb{R}$ can be extended to $\mathbb{C}$ preserving the postulate P_3. Then since $i \neq 0$, by P_1, we must have either $i > 0$ or $-i > 0$. If $i > 0$, by P_3, we get $i \cdot i > 0$, but this means $-1 > 0$, which is absurd. Similarly, if $-i > 0$, again by P_3, we get $(-i) \cdot (-i) > 0$, which implies again the absurd result $-1 > 0$. $\qquad \square$

Remark. Actually, there are infinitely many ways to define order relations in $\mathbb{C}$ satisfying P_1 and P_2. Here we have chosen a lexicographic one. As a consequence, *we should not use inequalities such as $<$ or $\geq$ for nonreal (complex) numbers.*

I.6 The Triangle Inequality

In §1.2, we defined the absolute value of a complex number $\alpha = a + ib$ $(a, b \in \mathbb{R})$ to be

$$|\alpha| = \sqrt{\alpha \overline{\alpha}} = \left(a^2 + b^2\right)^{1/2},$$

and mentioned some simple properties of the absolute value. We now prove the important *triangle inequality*.

THEOREM 1.6.1 (The Triangle Inequality). *For any $z_1, z_2 \in \mathbb{C}$,*

$$\big||z_1| - |z_2|\big| \leq |z_1 + z_2| \leq |z_1| + |z_2|.$$

Proof.

$$|z_1 + z_2|^2 = (z_1 + z_2)(\overline{z}_1 + \overline{z}_2)$$

$$= z_1 \overline{z}_1 + (z_1 \overline{z}_2 + \overline{z}_1 z_2) + z_2 \overline{z}_2$$

$$= |z_1|^2 + 2\Re(z_1 \overline{z}_2) + |z_2|^2$$

$$\leq |z_1|^2 + 2|z_1 \overline{z}_2| + |z_2|^2 \quad (\because \Re\alpha \leq |\alpha| \text{ for all } \alpha \in \mathbb{C})$$

$$= |z_1|^2 + 2|z_1| \cdot |z_2| + |z_2|^2 \quad (\because |\bar{z}_2| = |z_2|)$$

$$= (|z_1| + |z_2|)^2.$$

Since both $|z_1 + z_2|$ and $|z_1| + |z_2|$ are nonnegative, we obtain

$$|z_1 + z_2| \le |z_1| + |z_2|.$$

To prove the other inequality, note that $z_1 = (z_1 + z_2) + (-z_2)$.

$$\therefore \ |z_1| = |(z_1 + z_2) + (-z_2)| \le |z_1 + z_2| + |-z_2|$$

$$= |z_1 + z_2| + |z_2|.$$

It follows that

$$|z_1| - |z_2| \le |z_1 + z_2|.$$

Interchanging the roles of z_1 and z_2, we get

$$|z_2| - |z_1| \le |z_1 + z_2|.$$

$$\therefore \ \big||z_1| - |z_2|\big| \le |z_1 + z_2|.$$

□

We now discuss the situation where the triangle inequality becomes an equality. This is trivially the case if $z_1 = 0$ or $z_2 = 0$. Hence, we consider the case $z_1 \cdot z_2 \ne 0$. Looking back at the proof, we see that it becomes an equality if and only if

$$\Re(z_1 \bar{z}_2) = |z_1 \bar{z}_2|.$$

But a complex number α satisfies $\Re\alpha = |\alpha|$ if and only if α is a nonnegative real number. Hence the above inequality becomes an equality if and only if

$$z_1 \bar{z}_2 \ge 0.$$

Since we are assuming $z_2 \ne 0$, dividing both sides by $|z_2|^2 \ (= z_2 \bar{z}_2 > 0)$, we obtain that $\dfrac{z_1}{z_2} > 0$.

Summing up, $|z_1 + z_2| = |z_1| + |z_2|$ if and only if

$$\frac{z_1}{z_2} > 0 \qquad \text{unless} \quad z_1 = 0 \text{ or } z_2 = 0;$$

in other words, one of z_1 and z_2 has the property that the other is its positive multiple. We shall give another proof of this fact at the end of §1.8 Polar Representation of Complex Numbers. The reason for the name 'triangle inequality' will become clear in the next section.

1.7 The Complex Plane

We have defined a complex number to be an ordered pair of real numbers, but the set of all ordered pairs of real numbers has a one-to-one correspondence with the (x, y)-plane $\mathbb{R}^2$. So it is natural to let a complex number $z = x + iy$ correspond to the point (x, y) in the plane $\mathbb{R}^2$.

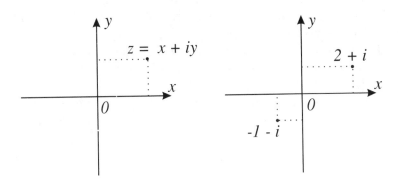

FIGURE 1.1

In the above correspondence, a real number $x = x + i0$ corresponds to the point $(x, 0)$ on the x-axis, and a purely imaginary number $iy = 0 + iy$ corresponds to the point $(0, y)$ on the y-axis, so we call the x-axis the *real axis*, and the y-axis the *imaginary axis*. A plane equipped with the real and imaginary axes is called the *complex plane* or the *Gaussian plane*.

Let us consider a sum of complex numbers in the complex plane $\mathbb{C}$. Suppose

$$z = x + iy, \qquad w = u + iv \qquad (x, y, u, v \in \mathbb{R}),$$

then

$$z + w = (x + u) + i(y + v).$$

This suggests that it is best to consider complex numbers as vectors; i.e., we regard a complex number $z = x + iy$ as a vector whose initial point is the origin and whose terminal point is the complex number $z = x + iy$. In other words, we consider a complex number $z = x + iy$ to be a vector whose orthogonal projections to the coordinate axes are x and y. Naturally, similar considerations apply to the complex number w. Then the sum $z + w$ corresponds to the diagonal vector (from the origin) of the parallelogram formed by two vectors z and w. Equivalently, draw the vector z with the initial point at the origin, and then draw the vector w with the initial point at the terminal point of z, then the vector with the initial point at the origin and the terminal point of w as its terminal point represents the vector $z + w$. (See Figure 1.2.)

From now on we shall identify a complex number with a point or a vector in the complex plane, whichever is most convenient for the particular situation.

We now consider the real multiple of a complex number. For $z = x + iy$ and $c \in \mathbb{R}$, we have $cz = cx + icy$. Thus, in the complex plane,

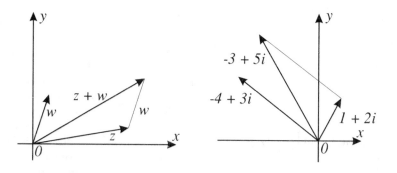

FIGURE 1.2

if $c > 0$, simply multiply the length of the vector by c (keeping the same direction), while if $c < 0$, multiply the length of the vector by $|c|$ and change to the opposite direction.

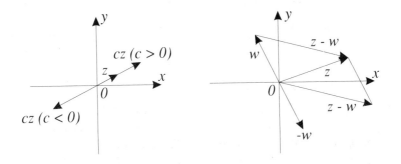

FIGURE 1.3

In particular, since $-w = (-1)w$, we obtain

$$z - w = z + (-w);$$

i.e., reverse the direction of the vector w to obtain $-w$, and draw the vector $-w$ with the initial point at the terminal point of z, then the vector with the initial point at that of z and the terminal point at that of $-w$ represents the vector $z - w$. Equivalently, $z - w$ is given by the vector with the initial point at the terminal point of w and the terminal point at that of z (provided that z and w have the same initial point). Note that $z + w$ and $z - w$ are two diagonals of a parallelogram with $\overrightarrow{0z}$ and $\overrightarrow{0w}$ as two neighboring sides.

EXAMPLE. Let z_1 and z_2 be two points in the complex plane. Then the midpoint of the line segment joining z_1 and z_2 is

$$z_1 + \frac{1}{2}(z_2 - z_1) = \frac{1}{2}(z_1 + z_2).$$

EXAMPLE. In an arbitrary $\triangle ABC$, let D and E be the midpoints of the sides AB and AC, respectively. Then DE is parallel to the side BC, and its length is half that of BC.

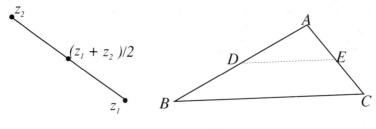

FIGURE I.4

Solution. Let $\triangle ABC$ be in the complex plane, and z_1, z_2, z_3 are the complex numbers representing the vertices A, B, C, respectively. Then the midpoints D and E of the sides AB and AC are given by

$$\frac{1}{2}(z_1 + z_2) \quad \text{and} \quad \frac{1}{2}(z_1 + z_3),$$

respectively. Hence the vector $\overrightarrow{DE}$ is given by

$$\frac{1}{2}(z_1 + z_3) - \frac{1}{2}(z_1 + z_2) = \frac{1}{2}(z_3 - z_2).$$

But $z_3 - z_2$ is precisely the vector $\overrightarrow{BC}$, hence the desired result.

EXAMPLE. The point z that divides the segment joining the points z_1 and z_2 into the ratio $m : n$ internally is given by

$$z = \frac{nz_1 + mz_2}{n + m},$$

where m and n are positive real numbers. For, it is easy to see that

$$\frac{z - z_1}{m} = \frac{z_2 - z}{n},$$

which gives the desired relation.

Equivalently, suppose z is an arbitrary point on the line segment joining the points z_1 and z_2, then

$$z - z_1 = t(z_2 - z_1) \quad \text{for some } t \in \mathbb{R} \quad (0 < t < 1);$$

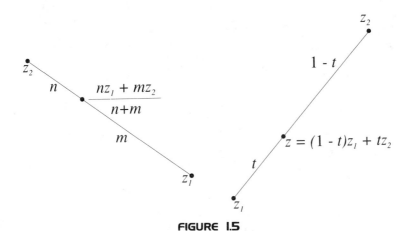

FIGURE I.5

$$\therefore z = (1 - t)z_1 + tz_2 \qquad (0 < t < 1).$$

Conversely, suppose this relation holds. Then since our argument is reversible, we can conclude that z must be a point in the line segment joining z_1 and z_2.

EXAMPLE. Let z_1, z_2, z_3 be three arbitrary points in the complex plane. Then the midpoint of the line segment joining z_2 and z_3 is $(z_2 + z_3)/2$, and so the parametric equation of the median through the vertex z_1 of $\triangle z_1 z_2 z_3$ is

$$z = (1 - t)z_1 + t \cdot \frac{z_2 + z_3}{2} \qquad (0 < t < 1).$$

Hence the point that divides this median into 2 : 1 internally, is obtained by substituting $t = \frac{2}{3}$ into the above expression; i.e.,

$$z = \frac{1}{3}(z_1 + z_2 + z_3).$$

But this expression is symmetric with respect to z_1, z_2, z_3, and hence this point also divides the medians from z_2 and from z_3 into 2 : 1 internally. Therefore, three medians of an arbitrary triangle intersect at a point. We call this point the *centroid* or the *center of gravity* of $\triangle z_1 z_2 z_3$.

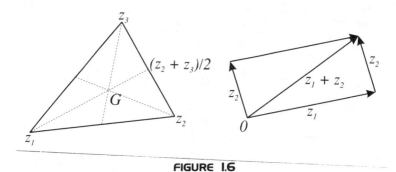

FIGURE 1.6

For $z = x + iy$ ($x, y \in \mathbb{R}$), we defined the absolute value of z to be

$$|z| = (z\bar{z})^{1/2} = \sqrt{x^2 + y^2};$$

i.e., $|z|$ is the length of the vector z. In other words, $|z|$ is the distance from the point z to the origin, which is in agreement with the case when z is real.

Note that $|z_1|$, $|z_2|$, $|z_1 + z_2|$ are the lengths of three sides of a triangle, and hence the name *triangle inequality* (Theorem 1.6.1):

$$|z_1 + z_2| \leq |z_1| + |z_2|.$$

It is geometrically obvious that the inequality becomes an equality only if the triangle degenerates to a line segment. We shall return to this point at the end of the next section.

1.8 Polar Representation of Complex Numbers

So far we have used only the vector aspect of complex numbers, hence we haven't seen the real power of complex numbers. Multiplication is well defined for complex numbers while it is not defined for vectors— the dot product (inner product) of two vectors is a scalar, not a vector, while the cross product of two vectors in a plane is a vector that is no longer in the plane. (The cross product is useful only in 3-dimensional space.) The essence of applications of complex numbers to plane geometry lies in the fact that the products of complex numbers are complex numbers.

For multiplication of complex numbers, it is convenient to use the polar representation of complex numbers. For a point $P = (x, y)$ ($= x + iy$) on a coordinate plane, consider the vector $\overrightarrow{OP}$ (where O is the origin). Let θ be the angle between $\overrightarrow{OP}$ and the positive x-axis, and $r = \overline{OP}$. Then $x = r\cos\theta$, $y = r\sin\theta$.

Naturally, θ is determined up to mod 2π; i.e., θ is determined uniquely if we neglect the difference of an integer multiple of 2π. The angle θ is called the *argument* of the complex number z.

Throughout this book, unless stated explicitly to the contrary, *equalities involving arguments will always be interpreted as congruence mod* 2π; *i.e., we shall neglect the difference of integer multiples of* 2π.

We call (r, θ) the *polar coordinates* of the point P.

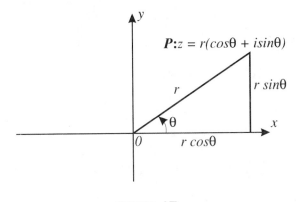

FIGURE I.7

The origin is the unique point where $r = 0$; the argument θ of the point at the origin is not defined.

If $z = x + iy$ ($\neq 0$), then we can write the polar representation of z:

$$z = r(\cos\theta + i\sin\theta).$$

That is, $r = |z|$ and $\theta = \arg z$ is the argument of z.

EXAMPLE.   For $z \in \mathbb{C}$, $z \neq 0$,

$$z \text{ is real} \iff \arg z = n\pi \qquad (n \in \mathbb{Z});$$

$$z \text{ is purely imaginary} \iff \arg z = \pm\frac{\pi}{2} + n\pi \qquad (n \in \mathbb{Z}).$$

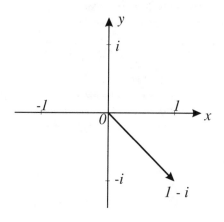

FIGURE I.8

EXAMPLE.

$$|-1| = 1, \qquad \arg(-1) = (2n + 1)\pi \quad (n \in \mathbb{Z});$$

$$|i| = 1, \qquad \arg(i) = \frac{\pi}{2} + 2n\pi \quad (n \in \mathbb{Z});$$

$$|1 - i| = \sqrt{1^2 + (-1)^2} = \sqrt{2}, \qquad \arg(1 - i) = -\frac{\pi}{4} + 2n\pi \quad (n \in \mathbb{Z}).$$

EXAMPLE.   Let $\omega = \dfrac{-1 + i\sqrt{3}}{2}$.
  Then

$$|\omega| = \sqrt{\left(-\frac{1}{2}\right)^2 + \left(\frac{\sqrt{3}}{2}\right)^2} = 1.$$

Set $\theta = \arg \omega$, then

$$\cos \theta = -\frac{1}{2}, \qquad \sin \theta = \frac{\sqrt{3}}{2},$$

and so $\theta = \frac{2\pi}{3} + 2n\pi$ $(n \in \mathbb{Z})$. Moreover,

$$|\overline{\omega}| = 1, \qquad \arg \overline{\omega} = -\frac{2\pi}{3} + 2n\pi.$$

Note that

$$\omega^2 = \overline{\omega}, \qquad \omega^2 + \omega + 1 = 0, \qquad \omega^3 = 1.$$

The points ω and ω^2 together with 1 form three vertices of an equilateral triangle inscribed in the unit circle.

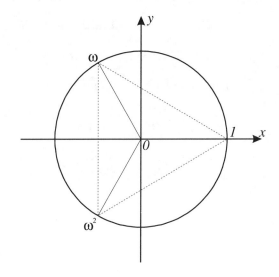

FIGURE 1.9

Polar representation is convenient for multiplication of complex numbers due to the following.

THEOREM 1.8.1. *Suppose*

$$z_1 = r_1(\cos\theta_1 + i\sin\theta_1), \qquad z_2 = r_2(\cos\theta_2 + i\sin\theta_2),$$

then

$$z_1 z_2 = r_1 r_2 \left\{ \cos(\theta_1 + \theta_2) + i\sin(\theta_1 + \theta_2) \right\};$$

i.e.,

$$|z_1 z_2| = |z_1| \cdot |z_2|, \qquad \arg(z_1 z_2) = \arg z_1 + \arg z_2.$$

In other words, the absolute value of the product is the product of the absolute values, and the argument of the product is the sum of the arguments.

Proof. By the addition formulas, we have

$$z_1 z_2 = r_1(\cos\theta_1 + i\sin\theta_1) \cdot r_2(\cos\theta_2 + i\sin\theta_2)$$

$$= r_1 r_2 \big\{ (\cos\theta_1 \cdot \cos\theta_2 - \sin\theta_1 \cdot \sin\theta_2)$$

$$+ i(\sin\theta_1 \cdot \cos\theta_2 + \cos\theta_1 \cdot \sin\theta_2) \big\}$$

$$= r_1 r_2 \big\{ \cos(\theta_1 + \theta_2) + i\sin(\theta_1 + \theta_2) \big\}.$$

$$\square$$

Remark. By restricting $\arg z$ to the interval $[0, 2\pi)$ or $(-\pi, \pi]$, $\arg z$ will be uniquely determined for all $z \in \mathbb{C}$ (except $z = 0$), but then the relation

$$\arg(z_1 z_2) = \arg z_1 + \arg z_2$$

will not be valid, and $\arg z$ will not be a continuous function of z.

COROLLARY 1.8.2.

$$|z_1 z_2 \cdots z_n| = |z_1| \cdot |z_2| \cdots |z_n|,$$

$$\arg(z_1 z_2 \cdots z_n) = \arg z_1 + \arg z_2 + \cdots + \arg z_n.$$

COROLLARY 1.8.3 (DeMoivre). *For $z = r(\cos\theta + i\sin\theta)$ and $n \in \mathbb{Z}$,*

$$z^n = r^n(\cos n\theta + i\sin n\theta),$$

viz.,

$$|z^n| = |z|^n, \qquad \arg(z^n) = n\arg(z).$$

Proof. We prove only the case $n = -1$.

$$(\cos\theta + i\sin\theta)(\cos\theta - i\sin\theta) = \cos^2\theta + \sin^2\theta = 1,$$

dividing both sides by $r(\cos\theta + i\sin\theta)$, we get

$$\frac{1}{r(\cos\theta + i\sin\theta)} = \frac{1}{r}(\cos\theta - i\sin\theta)$$

$$= \frac{1}{r}\{\cos(-\theta) + i\sin(-\theta)\},$$

since $\cos(-\theta) = \cos\theta$, $\sin(-\theta) = -\sin\theta$.

$$\therefore \ |z^{-1}| = |z|^{-1}, \qquad \arg\left(z^{-1}\right) = -\arg(z).$$

$\square$

COROLLARY 1.8.4.

$$\left|\frac{z_1}{z_2}\right| = \frac{|z_1|}{|z_2|}; \qquad \arg\frac{z_1}{z_2} = \arg z_1 - \arg z_2.$$

provided $z_2 \neq 0$.

EXAMPLE. From the DeMoivre formula, we can derive formulas for the sine and cosine functions. Choose $r = 1$, then it becomes

$$(\cos\theta + i\sin\theta)^n = \cos n\theta + i\sin n\theta.$$

In particular, for $n = 3$,

$$\cos 3\theta + i\sin 3\theta = (\cos\theta + i\sin\theta)^3$$

$$= \left(\cos^3 \theta - 3 \cos \theta \cdot \sin^2 \theta \right) + i \left(3 \cos^2 \theta \cdot \sin \theta - \sin^3 \theta \right).$$

$$\therefore \cos 3\theta = \cos^3 \theta - 3 \cos \theta \cdot \sin^2 \theta = 4 \cos^3 \theta - 3 \cos \theta,$$

$$\sin 3\theta = 3 \cos^2 \theta \cdot \sin \theta - \sin^3 \theta = 3 \sin \theta - 4 \sin^3 \theta.$$

Our theorem says that multiplying by z in the complex plane means magnifying (or contracting) a figure by the factor $|z|$, and rotating (counterclockwise) by the angle $\arg z$. In particular, multiplying by i means rotating by $\frac{\pi}{2}$ (counterclockwise).

With this preparation, given points z_1 and z_2 on the complex plane, we can construct the product $z_3 = z_1 z_2$ geometrically. All we need is an observation that $\triangle 01 z_1$ and $\triangle 0 z_2 z_3$ are similar (with the same orientation).

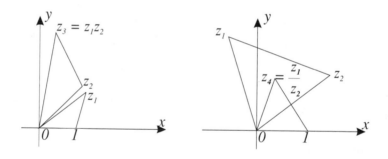

FIGURE 1.10

Similarly, to construct the quotient $z_4 = \frac{z_1}{z_2}$ geometrically, we need to observe that $\triangle 0 z_2 z_1$ and $\triangle 01 z_4$ are similar (with the same orientation).

EXAMPLE. Construct $1/(2 + i)$ geometrically. Let $z_1 = 2 + i$, $z_2 = 1/(2+i)$, then $\triangle 01 z_1$ and $\triangle 0 z_2 1$ are similar (with the same orientation), and so we construct z_2 as in Figure 1.11.

We now return to the case when the triangle inequality (Theorem 1.6.1),

$$|z_1 + z_2| \leq |z_1| + |z_2|,$$

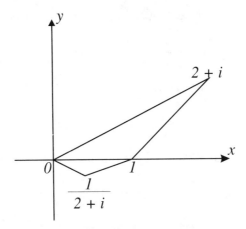

FIGURE I.II

becomes an equality. Examining the proof, we notice that equality holds if and only if any one of the following equivalent conditions holds:

1. $\Re(z_1 \bar{z}_2) = |z_1 \bar{z}_2|$.

2. $z_1 \bar{z}_2$ is a nonnegative real number.

3. $\dfrac{z_1}{z_2}$ is a positive real number or $z_1 z_2 = 0$.

4. z_1 and z_2 have the same argument (mod 2π).

5. z_1 and z_2 are on the same ray from the origin.

6. Vectors $\overrightarrow{0z_1}$ and $\overrightarrow{0z_2}$ have the same direction.

I.9   The nth Roots of 1

The unit circle—the circle with center at the origin and radius 1—can be expressed as

$$|\zeta| = 1 \qquad (\zeta \in \mathbb{C}).$$

In terms of polar representation, this can be written as

$$\zeta = \cos\theta + i\sin\theta \qquad (\theta \in \mathbb{R}).$$

For this ζ, if $z \in \mathbb{C}$, then we have

$$|\zeta z| = |\zeta| \cdot |z| = |z|, \qquad \arg(\zeta z) = \arg\zeta + \arg z.$$

It follows that multiplying by ζ simply means the rotation of z with respect to the origin by the angle θ ($= \arg\zeta$).

Suppose $z = x + iy, \zeta z = x' + iy'$. Then

$$x' + iy' = (\cos\theta + i\sin\theta)(x + iy)$$

$$= (x\cos\theta - y\sin\theta) + i(x\sin\theta + y\sin\theta).$$

$$\therefore \ x' = x\cos\theta - y\sin\theta,$$

$$y' = x\sin\theta + y\cos\theta.$$

This may remind us of a linear transformation. If we regard $x + iy$ and $x' + iy'$ as vectors $\binom{x}{y}$ and $\binom{x'}{y'}$, respectively, then the above relations can be written as

$$\begin{pmatrix} x' \\ y' \end{pmatrix} = \begin{pmatrix} \cos\theta & -\sin\theta \\ \sin\theta & \cos\theta \end{pmatrix} \begin{pmatrix} x \\ y \end{pmatrix}.$$

Conversely, if this relation holds, then

$$x' + iy' = (\cos\theta + i\sin\theta)(x + iy),$$

and so to multiply $\zeta = \cos\theta + i\sin\theta$ and $z = x + iy$ is the same as multiplying the rotation matrix

$$\begin{pmatrix} \cos\theta & -\sin\theta \\ \sin\theta & \cos\theta \end{pmatrix}$$

and the vector $\binom{x}{y}$.

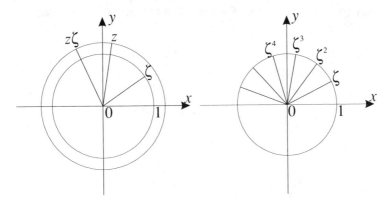

FIGURE 1.12

EXAMPLE. For $\zeta = \cos\theta + i\sin\theta$, ζ^2 is obtained by rotating ζ by the angle θ. Another rotation by θ gives ζ^3, etc.

EXAMPLE. For $\alpha = r(\cos\theta + i\sin\theta)$, α^2 is obtained by rotating α by the angle θ and magnifying (or contracting) its length by the factor r.

EXAMPLE. We now want to find all the cube roots of 1; i.e., we want to solve $z^3 = 1$.

Set $z = r(\cos\theta + i\sin\theta)$. Then, by the DeMoivre formula,

$$r^3(\cos 3\theta + i\sin 3\theta) = 1.$$

$$\therefore r^3 = 1, \qquad \cos 3\theta + i\sin 3\theta = 1.$$

Since r is a positive real number, we get

$$r = 1, \qquad \cos 3\theta = 1, \qquad \sin 3\theta = 0.$$

$$\therefore 3\theta = 2k\pi; \quad \text{i.e.,} \quad \theta = \frac{2k\pi}{3} \qquad (k = 0, \pm 1, \pm 2, \pm 3, \ldots).$$

According to the fundamental theorem of algebra (Theorem 1.4.1), there must be exactly 3 roots, yet it appears that we have found infinitely many roots. However, by the periodicity of the sine and cosine functions, we have

$$\omega_0 = 1 + i0 = \omega_3 = \omega_6 = \omega_9 = \cdots = \omega_{-3} = \omega_{-6} = \cdots,$$

$$\omega_1 = \cos\frac{2\pi}{3} + i\sin\frac{2\pi}{3} = \frac{-1 + i\sqrt{3}}{2}$$

$$= \omega_4 = \omega_7 = \omega_{10} = \cdots = \omega_{-2} = \omega_{-5} = \cdots,$$

$$\omega_2 = \cos\frac{4\pi}{3} + i\sin\frac{4\pi}{3} = \frac{-1 - i\sqrt{3}}{2}$$

$$= \omega_5 = \omega_8 = \omega_{11} = \cdots = \omega_{-1} = \omega_{-4} = \cdots.$$

These three points are the vertices of the equilateral triangle inscribed in the unit circle with one vertex at 1. Note that if we set $\omega = \omega_1$, then $\omega_2 = \omega^2 = \overline{\omega}$, $\omega^2 + \omega + 1 = 0$. (See Figure 1.13.)

We now embark on the task of finding all the solutions of the equation

$$z^n = 1.$$

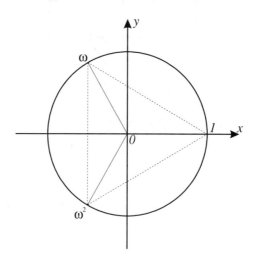

FIGURE 1.13

Let a solution be given in the polar form

$$z = r(\cos\theta + i\sin\theta).$$

By the DeMoivre formula,

$$z^n = r^n(\cos n\theta + i\sin n\theta).$$

$$\therefore r = 1, \qquad \cos n\theta + i\sin n\theta = 1.$$

Therefore,

$$n\theta = 2k\pi, \qquad \theta = \frac{2k\pi}{n} \qquad (k = 0, \pm 1, \pm 2, \ldots).$$

Conversely, it is simple to verify that

$$z_k = \cos\frac{2k\pi}{n} + i\sin\frac{2k\pi}{n} \qquad (k = 0, \pm 1, \pm 2, \ldots)$$

satisfy the given equation. We claim that

$$\text{if} \qquad k' \equiv k \pmod{n}, \qquad \text{then} \qquad z_{k'} = z_k.$$

Without loss of generality, we may assume that

$$k' = k + jn \qquad (j \in \mathbb{Z},\ 0 \le k < n).$$

Then

$$z_{k'} = \cos\frac{2k'\pi}{n} + i\sin\frac{2k'\pi}{n}$$

$$= \cos\left(\frac{2k\pi}{n} + 2j\pi\right) + i\sin\left(\frac{2k\pi}{n} + 2j\pi\right)$$

$$= \cos\frac{2k\pi}{n} + i\sin\frac{2k\pi}{n} = z_k,$$

so there are at most n distinct roots z_k corresponding to $k = 0, 1, \ldots, n - 1$. Conversely, if $z_{k'} = z_k$; that is, if

$$\cos \frac{2k'\pi}{n} + i \sin \frac{2k'\pi}{n} = \cos \frac{2k\pi}{n} + i \sin \frac{2k\pi}{n},$$

then we have

$$\frac{2k'\pi}{n} = \frac{2k\pi}{n} + 2m\pi \qquad \text{for some} \quad m \in \mathbb{Z},$$

which implies that $k' \equiv k \pmod{n}$.

Summing up, there are exactly n roots

$$\cos \frac{2k\pi}{n} + i \sin \frac{2k\pi}{n},$$

corresponding to $k = 0, 1, 2, \ldots, n - 1$. They form the vertices of the regular polygon with n sides inscribed in the unit circle with one of the vertices at 1.

EXAMPLE. Find the roots of $z^5 = 12 + 5i$.

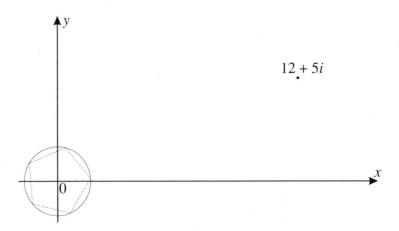

FIGURE I.I4

Solution. Let $\varphi = \arg(12 + 5i) = \arctan \frac{5}{12}$. Since $|12 + 5i| = 13$, if we set $z = r(\cos\theta + i\sin\theta)$, then

$$r^5(\cos 5\theta + i\sin 5\theta) = 13(\cos\varphi + i\sin\varphi).$$

$$\therefore r = \sqrt[5]{13}, \qquad \theta = \frac{\varphi}{5} + \frac{2k\pi}{5} \qquad (k = 0, 1, 2, 3, 4).$$

Thus, the five roots form vertices of the regular pentagon inscribed in a circle with the center at the origin and one vertex at $\sqrt[5]{13}(\cos\frac{\varphi}{5} + i\sin\frac{\varphi}{5})$ (see Figure 1.14). A reader should compare the polar representation approach with the method of §1.3 and §1.4.

EXAMPLE. Let $z = \cos\frac{2\pi}{5} + i\sin\frac{2\pi}{5}$. Then since $z^5 = 1$, but $z \neq 1$, we have

$$z^4 + z^3 + z^2 + z + 1 = 0.$$

Dividing both sides by z^2 $(z^2 \neq 0)$, we get

$$\left(z^2 + \frac{1}{z^2}\right) + \left(z + \frac{1}{z}\right) + 1 = 0;$$

i.e.,

$$\left(z + \frac{1}{z}\right)^2 + \left(z + \frac{1}{z}\right) - 1 = 0.$$

But $z + \frac{1}{z} = 2\cos\frac{2\pi}{5}$, so we get

$$4\cos^2\frac{2\pi}{5} + 2\cos\frac{2\pi}{5} - 1 = 0. \qquad \therefore \cos\frac{2\pi}{5} = \frac{-1 \pm \sqrt{5}}{4}.$$

But $\cos\frac{2\pi}{5} > 0$, so we obtain

$$\cos\frac{2\pi}{5} = \frac{\sqrt{5} - 1}{4}.$$

This result implies that *a regular pentagon can be constructed with a compass and straightedge.*

Quiz : Where does the other value $-\dfrac{\sqrt{5}+1}{4}$ come from? Is it a meaningless number simply to be discarded?

1.10 The Exponential function

In calculus[2] we learned that

$$e^x = 1 + \frac{x}{1!} + \frac{x^2}{2!} + \frac{x^3}{3!} + \cdots + \frac{x^n}{n!} + \cdots$$

is valid for all $x \in \mathbb{R}$. What if x is replaced by $i\theta$? The left-hand side becomes $e^{i\theta}$. But what is this exponential function with a complex variable? We don't know. So let us look at the right-hand side first. Every term $\dfrac{(i\theta)^n}{n!}$ makes perfectly good sense, so we collect the terms according to whether the term has i or not (we are changing the order of summation in an infinite series which is justified since our series is absolutely convergent), and we get

$$\left\{ 1 - \frac{\theta^2}{2!} + \frac{\theta^4}{4!} - \frac{\theta^6}{6!} + \cdots + (-1)^n \frac{\theta^{2n}}{(2n)!} + \cdots \right\}$$

$$+ i \left\{ \theta - \frac{\theta^3}{3!} + \frac{\theta^5}{5!} + \frac{\theta^7}{7!} + \cdots + (-1)^n \frac{\theta^{2n+1}}{(2n+1)!} + \cdots \right\}$$

$$= \cos\theta + i\sin\theta,$$

again by what we learned in calculus. Since $e^{i\theta}$ has no meaning, we might as well use this result to *define* it:

$$e^{i\theta} := \cos\theta + i\sin\theta.$$

---

[2] A reader who has no background in calculus may skip this section, treating $e^{i\theta}$ in what follows as a shorthand for $\cos\theta + i\sin\theta$, with the understanding that $e^{i\theta} \cdot e^{i\varphi} = e^{i(\theta+\varphi)}$, which is just a rewriting of the identity

$$(\cos\theta + i\sin\theta) \cdot (\cos\varphi + i\sin\varphi) = \cos(\theta + \varphi) + i\sin(\theta + \varphi).$$

Then the DeMoivre formula

$$(\cos\theta + i\sin\theta)^n = \cos n\theta + i\sin n\theta$$

reduces to

$$\left(e^{i\theta}\right)^n = e^{in\theta},$$

a triviality. Substituting $\theta = \pi$ in the equality defining $e^{i\theta}$, we get

$$e^{i\pi} = \cos\pi + i\sin\pi = -1, \qquad \therefore\ e^{i\pi} + 1 = 0,$$

which ties together the five numbers $0, 1, \pi, e, i$; arguably, these are the five most important numbers in mathematics.

Now the exponential function $f(x) = e^x$ $(x \in \mathbb{R})$ is characterized by the property

$$f(x + y) = f(x) \cdot f(y); \qquad \text{i.e.,} \qquad e^{x+y} = e^x \cdot e^y$$

(and the initial condition $f'(0) = 1$). Is this still valid when extended to a complex variable? The answer turns out to be affirmative, and there is a very slick proof that appeared in T. Takagi [*Mathematical Analysis*, Iwanami, Tokyo, 1986, p. 190], if we know that a complex power series can be differentiated termwise within the disc of convergence. Let

$$f(z) = 1 + \frac{z}{1!} + \frac{z^2}{2!} + \cdots + \frac{z^n}{n!} + \cdots.$$

By termwise differentiation, it is immediate that

$$f^{(n)}(z) = f(z) \qquad \text{for all} \qquad z \in \mathbb{C} \quad \text{and all} \quad n \in \mathbb{N}.$$

Therefore, by the Taylor series expansion of $f(z + w)$ around z, we get

$$f(z + w) = f(z) + \frac{f'(z)}{1!}w + \frac{f''(z)}{2!}w^2 + \cdots + \frac{f^{(n)}(z)}{n!}w^n + \cdots$$

$$= f(z)\left\{1 + \frac{w}{1!} + \frac{w^2}{2!} + \cdots + \frac{w^n}{n!} + \cdots\right\}$$

$$= f(z) \cdot f(w),$$

which is what we want to show.

Adding and subtracting the equalities

$$e^{i\theta} = \cos\theta + i\sin\theta$$

and

$$e^{-i\theta} = \cos\theta - i\sin\theta,$$

we get

$$\cos\theta = \frac{e^{i\theta} + e^{-i\theta}}{2}, \qquad \sin\theta = \frac{e^{i\theta} - e^{-i\theta}}{2i}.$$

Multiplying, we get

$$\cos^2\theta + \sin^2\theta = 1.$$

EXAMPLE.   Show that

$$\frac{d^n}{dx^n}\left(e^x\cos x\right) = 2^{n/2}e^x\cos\left(x + \frac{n\pi}{4}\right) \qquad (n = 0, 1, 2, \ldots).$$

Solution.   This can be done by mathematical induction, but the following computation gives insight, and allows generalization.

$$\frac{d^n}{dx^n}\left(e^x\cos x\right) = \frac{d^n}{dx^n}\left\{\Re e^{(1+i)x}\right\}$$

$$= \Re\left\{\frac{d^n}{dx^n}e^{(1+i)x}\right\}$$

$$= \Re\left\{(1 + i)^n \cdot e^{(1+i)x}\right\}$$

$$= \Re\left\{\left(\sqrt{2} \cdot e^{i(\pi/4)}\right)^n \cdot e^{(1+i)x}\right\}$$

$$= 2^{n/2} \cdot \Re\left\{e^x \cdot e^{i(x+(n\pi/4))}\right\}$$

$$= 2^{n/2} \cdot e^x \cdot \cos\left(x + \frac{n\pi}{4}\right).$$

EXAMPLE.   Show that for $0 \leq r < 1$,

$$1 + 2\left(r\cos\theta + r^2\cos 2\theta + \cdots + r^n\cos n\theta + \cdots\right) = \frac{1-r^2}{1-2r\cos\theta + r^2}.$$

Solution.

$$1 + 2\sum_{n=1}^{\infty} r^n\cos n\theta = 1 + 2\sum_{n=1}^{\infty}\Re\left(r^n e^{in\theta}\right)$$

$$= 1 + 2\Re\left\{\sum_{n=1}^{\infty}\left(re^{i\theta}\right)^n\right\}$$

$$= 1 + 2\Re\left\{\frac{re^{i\theta}}{1-re^{i\theta}}\right\}$$

$$= 1 + 2\frac{\Re\left\{re^{i\theta}\left(1-re^{-i\theta}\right)\right\}}{(1-re^{i\theta})(1-re^{-i\theta})}$$

$$= 1 + \frac{2\left(r\cos\theta - r^2\right)}{1-2r\cos\theta + r^2}$$

$$= \frac{1-r^2}{1-2r\cos\theta + r^2},$$

which is the important *Poisson kernel.* Taking the imaginary part instead of the real part in the above computation, we obtain the *conjugate Poisson kernel:*

$$\sum_{n=1}^{\infty} r^n\sin n\theta = \frac{r\sin\theta}{1-2r\cos\theta + r^2}\qquad (0 \leq r < 1).$$

EXAMPLE.   It is simple to verify that

$$\int_0^{2\pi} e^{in\theta}\,d\theta = \begin{cases} 2\pi & (n=0); \\ 0 & (n=\pm1,\ \pm2,\ \pm3,\cdots). \end{cases}$$

$$\therefore \int_0^{2\pi} \cos^4 \theta \, d\theta = \int_0^{2\pi} \left(\frac{e^{i\theta} + e^{-i\theta}}{2} \right)^4 d\theta$$

$$= \frac{1}{2^4} \int_0^{2\pi} \left(e^{4i\theta} + 4e^{2i\theta} + 6 + 4e^{-2i\theta} + e^{-4i\theta} \right) d\theta$$

$$= \frac{6}{16} \int_0^{2\pi} d\theta = \frac{3\pi}{4}.$$

More generally, for $n \in \mathbb{N}$, we have

$$\int_0^{2\pi} \sin^{2n} \theta \, d\theta = \int_0^{2\pi} \cos^{2n} \theta \, d\theta = \binom{2n}{n} \cdot \frac{2\pi}{2^{2n}}$$

$$= \frac{(2n)!}{(n!)^2} \cdot \frac{2\pi}{2^{2n}}$$

$$= \frac{1 \cdot 3 \cdot 5 \cdots (2n-3) \cdot (2n-1)}{2 \cdot 4 \cdot 6 \cdots (2n-2) \cdot (2n)} \cdot 2\pi \quad \text{(Wallis)}.$$

Exercises

1. Perform the indicated operations, and reduce each of the following numbers to the form $x + iy$ ($x, y \in \mathbb{R}$):

   (a) $(1 - i)(2 - i)(3 - i)$;

   (b) $\left(\sqrt{3} + i \right)^6$;

   (c) $\dfrac{4 + 3i}{3 - 4i}$;

   (d) $\dfrac{5 - z}{5 + z}$, where $z = 4 + 3i$.

2. Find the real numbers x, y, u, v satisfying

$$z = x + i, \qquad w = 3 + iy,$$

$$z + w = u - i, \qquad zw = 14 + iv.$$

3. Let $z = a + ib \, (a, b \in \mathbb{R})$.

   (a) Express $|\Re(z^2)|^2 + |\Im(z^2)|^2 = |z|^4$ in terms of a and b.

   (b) If $z = 2 + i$, then this gives us $3^2 + 4^2 = 5^2$. Can you find other *Pythagorean triples*?

4. Show that any complex number z with $|z| = 1$, but $z \neq -1$ can be expressed as

$$z = \frac{1 + it}{1 - it}$$

   with an appropriate choice of the real parameter t.

5. Let $z = a + ib \, (a, b \in \mathbb{R})$. Find conditions on a and b such that

   (a) z^4 is real;

   (b) z^4 is purely imaginary.

6. Find the absolute values of

   (a) $3 + 2i$;

   (b) $-1 + i\sqrt{3}$;

   (c) $-i(1 + i)(2 - 3i)(4 + 3i)$;

   (d) $\dfrac{(3 - i)(-1 + 2i)}{2 - 3i}$.

7. Let $z = a + ib, \, w = c + id \, (a, b, c, d \in \mathbb{R})$.

   (a) Express $|zw| = |z||w| = |z\overline{w}|$ in terms of a, b, c, d.

   (b) Choosing $z = 2 + i, \, w = 2 + 3i$ in (a), we get $8^2 + 1 = 4^2 + 7^2$. Find other positive integers $p, q, r \, (p \neq 1, q \neq 1)$ satisfying the equation

$$r^2 + 1 = p^2 + q^2.$$

   (c) Show that the set

$$S = \{p \in \mathbb{N} \, ; \, p = m^2 + n^2 \text{ for some } m, \, n \in \mathbb{N}\}$$

is closed under multiplication; i.e., $p, q \in S \implies pq \in S$.

8. (a) Prove the *parallelogram law*:

$$|\alpha + \beta|^2 + |\alpha - \beta|^2 = 2(|\alpha|^2 + |\beta|^2)$$

for two arbitrary complex numbers α and β.

(b) Suppose that $|\alpha| = |\beta|$. Show that for any $\gamma \in \mathbb{C}$,

$$|\alpha + \gamma|^2 + |\alpha - \gamma|^2 = |\beta + \gamma|^2 + |\beta - \gamma|^2.$$

(c) Interpret the equalities in (a) and (b) geometrically.

9. If $|\alpha| < 1$ and $|z| \leq 1$, show that $\left| \frac{z + \alpha}{1 + \bar{\alpha}z} \right| \leq 1$. When does equality hold?

10. Find the set of all $z \in \mathbb{C}$ satisfying the following :

(a) $\Re z \geq \Im z$;

(b) $|z - 1 + 3i| < 4$;

(c) $|z - 1| + |z + i| = 2$.

(d) $|z - 1 + i| - |z + 1 - i| \geq 2$.

11. Let $|\alpha| = |\beta| = |\gamma| = 1$.

(a) Suppose $\alpha + \beta + \gamma \neq 0$. Show that $\left| \frac{\beta\gamma + \gamma\alpha + \alpha\beta}{\alpha + \beta + \gamma} \right| = 1$.

(b) Show that $\frac{(\beta + \gamma)(\gamma + \alpha)(\alpha + \beta)}{\alpha\beta\gamma} \in \mathbb{R}$.

12. Solve the following quadratic equations :

(a) $\frac{1}{2}z^2 + (1 - i)z + i = 0$. (Compare with the example in §1.3. What conclusion can you draw?)

(b) $(1 - i)z^2 - 3z - (1 + i) = 0$. (What is the discriminant? Are the roots real?)

13. (a) Show that if α is a root of a polynomial equation with real coefficients (i.e., all the coefficients are real), then $\overline{\alpha}$ is also a root.

(b) Show that a polynomial equation with real coefficients and odd degree must have at least one real root.

14. What is wrong with the following 'proof'?

$$1 = \sqrt{1^2} = \sqrt{(-1)^2} = \left(\sqrt{-1}\right)^2 = i^2 = -1. \qquad \therefore 2 = 0.$$

15. Solve

(a) $z^3 - i = 0$;

(b) $z^4 + 1 = 0$;

(c) $z^5 + 32 = 0$;

(d) $z^6 - 1 = 0$.

16. Give another example of an order relation in $\mathbb{C}$ satisfying the postulates P_1 and P_2 in §1.5.

17. Let $a, b, c \in \mathbb{R}$. Prove

(a) $a > b,\ b > c \implies a > c$;

(b) $a > b,\ c > 0 \implies ac > bc$;

(c) $a > b,\ c < 0 \implies ac < bc$,

from the postulates P_1, P_2, and P_3 in §1.5.

18. If $\zeta^5 = 1$, show that

(a) $\dfrac{\zeta}{1 + \zeta^2} + \dfrac{\zeta^2}{1 + \zeta^4} + \dfrac{\zeta^3}{1 + \zeta} + \dfrac{\zeta^4}{1 + \zeta^3} = 2$;

(b) $\dfrac{\zeta}{1 - \zeta^2} + \dfrac{\zeta^2}{1 - \zeta^4} + \dfrac{\zeta^3}{1 - \zeta} + \dfrac{\zeta^4}{1 - \zeta^3} = 0\ (\zeta \neq 1)$.

19. Let $\omega^2 + \omega + 1 = 0$.

   (a) Show that every complex number $z \in \mathbb{C}$ can be expressed
   uniquely in the form

   $$z = a + b\omega \qquad (a, b \in \mathbb{R}).$$

   (b) Find a and b $(a, b \in \mathbb{R})$ such that $\dfrac{7 + 5\omega + 3\omega^2}{1 - 2\omega} = a + b\omega$.

20. Show that for arbitrary $z_1, z_2, \ldots, z_n \in \mathbb{C}$, we have

   $$|z_1 + z_2 + \cdots + z_n| \le |z_1| + |z_2| + \cdots + |z_n|.$$

   When does equality hold?

21. Given three vertices $3 + i$, $1 - 2i$, $-2 + 4i$ of a parallelogram, find
   the fourth vertex. How many solutions are there?

22. Show that the diagonals of a parallelogram bisect each other.

23. Show that in an arbitrary quadrangle, the midpoints of the four
   sides are the vertices of a parallelogram.

24. We have seen in §1.7, that a point z is on the line segment joining
   the points z_1 and z_2 if and only if

   $$z = (1 - t)z_1 + tz_2 \qquad \text{for some} \quad t \in (0, 1).$$

   What if $t \in \mathbb{R}$ is not in this range?

25. (a) Show that z_1, z_2, $z_3 \in \mathbb{C}$ are collinear if and only if there are
   three real numbers α, β, γ, not all zero, such that

   $$\alpha z_1 + \beta z_2 + \gamma z_3 = 0, \qquad \alpha + \beta + \gamma = 0.$$

   (b) Can this be extended to four points or more?

26. Given a hexagon, if we choose the midpoints of alternate sides, we
   obtain the vertices of two triangles. Show that the centroids of these
   two triangles coincide.

27. Let A', B', C' be points on (the extensions of) the respective sides BC, CA, AB of $\triangle ABC$ such that

$$\frac{\overline{BA'}}{\overline{BC}} = \frac{\overline{CB'}}{\overline{CA}} = \frac{\overline{AC'}}{\overline{AB}},$$

where the orientations of the line segments are taken into account; i.e., $\frac{\overline{BA'}}{\overline{BC}} > 0$ if $\overrightarrow{BA'}$ and $\overrightarrow{BC}$ are of the same direction, and $\frac{\overline{BA'}}{\overline{BC}} < 0$ if they are of the opposite direction; and similarly for the other ratios. Show that the centroids of $\triangle A'B'C'$ and $\triangle ABC$ coincide.

28. (a) Given four arbitrary points z_1, z_2, z_3, z_4 in the complex plane, let w_1, w_2, w_3, w_4 be the centroids of

$$\triangle z_2 z_3 z_4, \quad \triangle z_1 z_3 z_4, \quad \triangle z_1 z_2 z_4, \quad \triangle z_1 z_2 z_3,$$

respectively. (If you so wish, you may assume that no three of these four points are collinear.) Show that the four line segments joining the points z_1 and w_1, z_2 and w_2, z_3 and w_3, z_4 and w_4 intersect at one point.

(b) Generalize.

29. (a) Given 3 points, construct a triangle for which these points are the midpoints of the sides.

(b) Given 5 points, construct a pentagon (may be self-intersecting) for which these points are the midpoints of the sides.

(c) Generalize.

(d) What if the number of the given points is even?

30. Let z_1, z_2, z_3 be three arbitrary points in the complex plane.

(a) Show that a point z is in the interior or on the boundary of $\triangle z_1 z_2 z_3$ if and only if there are nonnegative real numbers α, β, γ such that

$$z = \alpha z_1 + \beta z_2 + \gamma z_3, \qquad \alpha + \beta + \gamma = 1.$$

(b) Find the locus of the points for which $\alpha = \frac{1}{3}$.

(c) Show also that if the triangle $\triangle z_1 z_2 z_3$ does not degenerate, then the correspondence between the set of points z in the (closed) triangle (i.e., the boundary included) and the set of ordered triples

$$\left\{(\alpha, \beta, \gamma) \in \mathbb{R}^3;\ \alpha \geq 0,\ \beta \geq 0,\ \gamma \geq 0,\ \alpha + \beta + \gamma = 1\right\}$$

is one-to-one. [$(\alpha,\ \beta,\ \gamma)$ are called the *barycentric coordinates* of the point z.]

31. (a) Let $z_1, z_2, \ldots, z_n$ be arbitrary points in the complex plane ($n \geq 2$). Show that a point z is in the smallest closed convex polygon containing these n points if and only if there are nonnegative real numbers $\alpha_1, \alpha_2, \ldots, \alpha_n$ such that

$$z = \sum_{j=1}^{n} \alpha_j z_j, \qquad \sum_{j=1}^{n} \alpha_j = 1.$$

(b) Assuming $z_1, z_2, \ldots, z_n$ are vertices of a convex n-gon, do we have uniqueness of representation as in the previous problem?

32. (a) Suppose $0 \leq \arg w - \arg z < \pi$, show that the area of $\triangle 0zw$ is given by

$$\frac{1}{2} \Im(\bar{z}w).$$

(b) Suppose $0 \leq \arg z_1 < \arg z_2 < \cdots < \arg z_n < 2\pi$, show that the area of the polygon whose vertices are at $z_1, z_2, \ldots, z_n$ is given by

$$\frac{1}{2} \Im \left\{ \sum_{k=1}^{n} \bar{z}_{k-1} z_k \right\} \qquad (z_0 = z_n).$$

(c) Show that the result in (b) can also be written as

$$\frac{1}{4i} \sum_{k=1}^{n} (z_k - z_{k-1})(\bar{z}_k + \bar{z}_{k-1}).$$

33. (a) Given $z \in \mathbb{C}$, show that there exist $\alpha, \beta \in \mathbb{C}$ with $|\alpha| = |\beta| = 1$ such that $z = \alpha + \beta$ if and only if $|z| \leq 2$.

   (b) Given $z \in \mathbb{C}$, show that there exist $\alpha, \beta, \gamma \in \mathbb{C}$ with $|\alpha| = |\beta| = |\gamma| = 1$ such that $z = \alpha + \beta + \gamma$ if and only if $|z| \leq 3$.

   (c) Generalize.

34. (a) Find the condition on $a, b \in \mathbb{R}$ for which the system of simultaneous equations

$$\cos x + \cos y = a, \qquad \sin x + \sin y = b,$$

   has solutions $x, y \in \mathbb{R}$.

   (b) Solve the above system of equations for the case $a = \dfrac{\sqrt{2}}{2}$, $b = \dfrac{\sqrt{6}}{2}$.

   (c) Solve $\cos x + \sin y = \dfrac{\sqrt{6}}{2}$, $\sin x + \cos y = -\dfrac{\sqrt{6}}{2}$.

   (d) Solve $5 \cos x + 3 \sin y = -\dfrac{7\sqrt{2}}{2}$, $5 \sin x - 3 \cos y = \dfrac{7\sqrt{2}}{2}$.

35. (a) If $z_1 + z_2 + z_3 = 0$ and $|z_1| = |z_2| = |z_3| = 1$, show that z_1, z_2, z_3 are the vertices of an equilateral triangle inscribed in the unit circle.

   (b) If $z_1 + z_2 + z_3 + z_4 = 0$ and $|z_1| = |z_2| = |z_3| = |z_4| = 1$, what can be said about the quadrangle with vertices at z_1, z_2, z_3, z_4?

36. For any complex number $a \neq 0$, show that a, $-a$, $\dfrac{1}{a}$, $-\dfrac{1}{a}$, 0 are collinear.

37. Find polar representations for

   (a) $1 + i$;

   (b) $4 - 3i$;

   (c) $1 + \omega$;

(d) $\dfrac{1}{\omega}$,

where $\omega^2 + \omega + 1 = 0$.

38. Give a counterexample to

$$\arg(z_1 z_2) = \arg z_1 + \arg z_2$$

if we make the restriction $0 \le \arg z < 2\pi$. What if we make the restriction $-\pi < \arg z \le \pi$?

39. (a) Show that $1 + z + z^2 + \cdots + z^n = \dfrac{1 - z^{n+1}}{1 - z}$ $(z \ne 1)$.

(b) Suppose $\zeta^{17} = 1$ $(\zeta \ne 1)$. Show that

$$1 + \zeta^k + \zeta^{2k} + \cdots + \zeta^{16k} = 0$$

where k is an arbitrary integer that is not a multiple of 17.

40. Let θ be the exterior angle of a regular n-gon. Show that

(a) $1 + \cos\theta + \cos 2\theta + \cdots + \cos(n-1)\theta = 0$;

(b) $\sin\theta + \sin 2\theta + \cdots + \sin(n-1)\theta = 0$.

41. (a) Derive the following identities :

$$\sum_{k=0}^{n} \cos k\theta = \frac{\cos \frac{n\theta}{2} \cdot \sin \frac{(n+1)\theta}{2}}{\sin \frac{\theta}{2}},$$

$$\sum_{k=1}^{n} \sin k\theta = \frac{\sin \frac{n\theta}{2} \cdot \sin \frac{(n+1)\theta}{2}}{\sin \frac{\theta}{2}} \qquad (0 < \theta < 2\pi).$$

(b) Use the identities in (a) to evaluate the sums:

$$\sum_{k=1}^{n} k = 1 + 2 + 3 + \cdots + n,$$

$$\sum_{k=1}^{n} k^2 = 1^2 + 2^2 + 3^2 + \cdots + n^2.$$

[This serves to check our results in (a).]

Hint: $\lim\limits_{\theta \to 0} \dfrac{\sin k\theta}{\theta} = ?$ $\lim\limits_{\theta \to 0} \dfrac{1 - \cos k\theta}{\theta^2} = ?$

42. Show that

$$\binom{n}{0} - \binom{n}{2} + \binom{n}{4} - + \cdots = 2^{n/2} \cos \frac{n\pi}{4},$$

$$\binom{n}{1} - \binom{n}{3} + \binom{n}{5} - + \cdots = 2^{n/2} \sin \frac{n\pi}{4}.$$

Hint : $(1 + z)^n = \binom{n}{0} + \binom{n}{1} z + \cdots + \binom{n}{n} z^n.$

43. (a) Express $1 + i\sqrt{3}$ in polar form.

(b) Simplify $\left(1 + i\sqrt{3}\right)^{1991} + \left(1 - i\sqrt{3}\right)^{1991}.$

(c) Simplify $(1 + i)^{1991} + (1 - i)^{1991}, (1 + i)^{1991} - (1 - i)^{1991}.$

44. Suppose $x_n + iy_n = \left(1 + i\sqrt{3}\right)^n$ $(x_n, y_n \in \mathbb{R})$. Show that

$$x_n y_{n+1} - x_{n+1} y_n = 2^{2n}\sqrt{3},$$

$$x_{n+1} x_n + y_{n+1} y_n = 2^{2n}.$$

45. Find the smallest positive integers m and n satisfying

$$\left(1 + i\sqrt{3}\right)^m = (1 - i)^n.$$

46. Let $x = a + b, y = a\omega + b\omega^2, z = a\omega^2 + b\omega$, where $\omega^2 + \omega + 1 = 0$. Express $x^3 + y^3 + z^3$ in terms of a and b.

47. Let $\omega^2 + \omega + 1 = 0$.

(a) Given any two polynomials $p(z)$ and $q(z)$, show that the polynomials

$$f(z) = p(z)p(\omega z)p(\omega^2 z),$$

$$g(z) = p(\omega z)q(\omega^2 z) + p(\omega^2 z)q(z) + p(z)q(\omega z)$$

have nonzero coefficients a_k, b_k only if k is a multiple of 3, where a_k, b_k are the coefficients of z^k in $f(z)$ and $g(z)$, respectively.

(b) Prove that every function $\varphi(z)$ (defined for all $z \in \mathbb{C}$) can be expressed as

$$\varphi(z) = f(z) + g(z) + h(z),$$

where $f(\omega z) = f(z)$, $g(\omega z) = \omega g(z)$, $h(\omega z) = \omega^2 h(z)$ for all $z \in \mathbb{C}$.

(c) Generalize.

48. Show that

(a) $\sin 2\theta = 2 \sin \theta \cdot \cos \theta$,

$$\sin 3\theta = 4 \sin \theta \cdot \sin \left(\frac{\pi}{3} - \theta \right) \cdot \sin \left(\frac{\pi}{3} + \theta \right),$$

$$\sin 4\theta = 8 \sin \theta \cdot \cos \theta \cdot \sin \left(\frac{\pi}{4} - \theta \right) \cdot \sin \left(\frac{\pi}{4} + \theta \right),$$

$$\sin 5\theta = 16 \sin \theta \cdot \sin \left(\frac{\pi}{5} - \theta \right) \cdot \sin \left(\frac{2\pi}{5} - \theta \right)$$

$$\cdot \sin \left(\frac{\pi}{5} + \theta \right) \cdot \sin \left(\frac{2\pi}{5} + \theta \right).$$

(b) $\cos 2\theta = 2 \sin \left(\frac{\pi}{4} - \theta \right) \cdot \sin \left(\frac{\pi}{4} + \theta \right),$

$$\cos 3\theta = 4 \cos \theta \cdot \sin \left(\frac{\pi}{6} - \theta \right) \cdot \sin \left(\frac{\pi}{6} + \theta \right),$$

$$\cos 4\theta = 8 \sin \left(\frac{\pi}{8} - \theta \right) \cdot \sin \left(\frac{3\pi}{8} - \theta \right) \cdot \sin \left(\frac{\pi}{8} + \theta \right)$$
$$\cdot \sin \left(\frac{3\pi}{8} + \theta \right),$$

$$\cos 5\theta = 16 \cos \theta \cdot \sin \left(\frac{\pi}{10} - \theta \right) \cdot \sin \left(\frac{3\pi}{10} - \theta \right)$$
$$\cdot \sin \left(\frac{\pi}{10} + \theta \right) \cdot \sin \left(\frac{3\pi}{10} + \theta \right).$$

(c) Generalize.

49. (a) Let $\zeta = e^{2\pi i/n}$, show that

$$\prod_{k=1}^{n} (z - \zeta^k) = z^n - 1,$$

$$\prod_{k=1}^{n-1} (z - \zeta^k) = 1 + z + \cdots + z^{n-1}.$$

(b) Show that

$$\sin \frac{\pi}{n} \cdot \sin \frac{2\pi}{n} \cdots \sin \frac{(n-1)\pi}{n} = \frac{n}{2^{n-1}} \qquad (n \geq 2).$$

50. Construct a regular pentagon using only a compass and straight-edge.

51. Suppose $z + \dfrac{1}{z} = 1$. Show that the sequence $\{w_k\}_{k=0}^{\infty}$, where $w_k = z^k + \dfrac{1}{z^k}$, is periodic, and find its period.

52. (a) Show that the polynomial $z^{2n} + z^n + 1$ ($n \in \mathbb{N}$) is divisible by $z^2 + z + 1$ if and only if n is not a multiple of 3.

(b) Find a necessary and sufficient condition on natural numbers p and q such that the polynomial $z^p + z^q + 1$ ($n \in \mathbb{N}$) is divisible by $z^2 + z + 1$.

53. Show that[3]

    (a) For every natural number n, there exist polynomials $p_n(x)$ and $q_n(x)$ (with real coefficients) satisfying

$$\cos n\theta = p_n(\tan \theta) \cdot \cos^n \theta,$$

$$\sin n\theta = q_n(\tan \theta) \cdot \cos^n \theta.$$

    (b) $p_n(x) = \frac{1}{2}\{(1 + ix)^n + (1 - ix)^n\}$,
       $q_n(x) = \frac{1}{2i}\{(1 + ix)^n - (1 - ix)^n\}$.

    (c) $p_n'(x) = -nq_{n-1}(x)$, $q_n'(x) = np_{n-1}(x)$ $(n > 1)$.

54. Guess and prove similar relations to the Example in §1.10 :

    (a) $\dfrac{d^n}{dx^n}(e^{-x} \cdot \sin x)$;

    (b) $\dfrac{d^n}{dx^n}(e^{\sqrt{3}x} \cdot \cos x)$;

    (c) $\dfrac{d^n}{dx^n}(e^{-x} \cdot \sin \sqrt{3}x)$.

55. Evaluate

    (a) $H = \displaystyle\int e^{ax} \cdot \cos bx \, dx$,

    (b) $K = \displaystyle\int e^{ax} \cdot \sin bx \, dx$.

    Hint: $H + iK = ?$

---

[3] Problem 53, excluding part (b), is taken from the entrance examination of the University of Tokyo, Japan, February 25, 1991.

CHAPTER 2

Applications to Geometry

2.1 Triangles

We now discuss applications of complex numbers to plane geometry. It is important to keep in mind that complex numbers are not just vectors; they can be multiplied by each other. In applications to geometry, we shall make full use of this property. Complex numbers are particularly effective for certain types of problems, but may be cumbersome for some problems that can be solved by elementary methods.

In elementary geometry, triangles are the building blocks and the congruence and similarity of two triangles are the most fundamental concepts. We start from the conditions on the similarity of two triangles in terms of complex numbers. Let us first present the following notational conventions and some review. Throughout this chapter, we say $\triangle z_1 z_2 z_3$ and $\triangle w_1 w_2 w_3$ are *similar*, and write

$$\triangle z_1 z_2 z_3 \sim \triangle w_1 w_2 w_3$$

if and only if the angle at z_k is equal to that at w_k (hence z_k correspond to w_k, $k = 1, 2, 3$), *and* they are of the same orientation (i.e., they are both counterclockwise or both clockwise).

If they are of the opposite orientation (one clockwise, the other counterclockwise), then we write

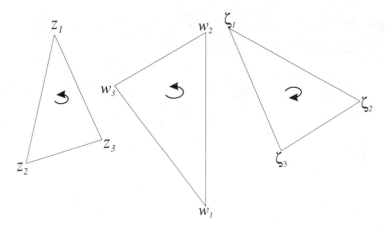

FIGURE 2.1

$$\triangle z_1 z_2 z_3 \sim \triangle w_1 w_2 w_3 \qquad \text{(reversed).}$$

Note that z_k still must correspond to w_k ($k = 1, 2, 3$).

As usual, we use the notations $\|$, $\perp$ to denote that two lines (line segments or vectors) are parallel or orthogonal, respectively.

Since for distinct points $\alpha, \beta, \gamma \in \mathbb{C}$,

$$\arg \frac{\beta - \alpha}{\gamma - \alpha} = \arg(\beta - \alpha) - \arg(\gamma - \alpha)$$

$$= \text{the oriented angle from the vector } \overrightarrow{\alpha\gamma} \text{ to } \overrightarrow{\alpha\beta},$$

$$\alpha, \ \beta, \ \gamma \text{ are collinear} \iff \frac{\beta - \alpha}{\gamma - \alpha} \in \mathbb{R}$$

$$\iff \frac{\beta - \alpha}{\gamma - \alpha} = \frac{\overline{\beta} - \overline{\alpha}}{\overline{\gamma} - \overline{\alpha}};$$

and

$$\overrightarrow{\alpha\beta} \perp \overrightarrow{\alpha\gamma} \iff \frac{\beta - \alpha}{\gamma - \alpha} \text{ is purely imaginary}$$

$$\Longleftrightarrow \frac{\beta - \alpha}{\gamma - \alpha} + \frac{\overline{\beta} - \overline{\alpha}}{\overline{\gamma} - \overline{\alpha}} = 0.$$

More generally, for four distinct points $\alpha, \beta, \gamma, \delta \in \mathbb{C}$,

$$\overrightarrow{\alpha\beta} \parallel \overrightarrow{\gamma\delta} \Longleftrightarrow \frac{\beta - \alpha}{\delta - \gamma} \in \mathbb{R}$$

$$\Longleftrightarrow \frac{\beta - \alpha}{\delta - \gamma} = \frac{\overline{\beta} - \overline{\alpha}}{\overline{\delta} - \overline{\gamma}};$$

furthermore, $\overrightarrow{\alpha\beta}$ and $\overrightarrow{\gamma\delta}$ have the same (opposite) direction if and only if $\frac{\beta - \alpha}{\delta - \gamma}$ is a positive (negative) real number; and

$$\overrightarrow{\alpha\beta} \perp \overrightarrow{\gamma\delta} \Longleftrightarrow \frac{\beta - \alpha}{\delta - \gamma} \quad \text{is purely imaginary}$$

$$\Longleftrightarrow \frac{\beta - \alpha}{\delta - \gamma} + \frac{\overline{\beta} - \overline{\alpha}}{\overline{\delta} - \overline{\gamma}} = 0.$$

Consequently, if $\alpha\beta \neq 0$, then

$$|\alpha + \beta| = |\alpha| + |\beta| \Longleftrightarrow \frac{\alpha}{\beta} \in \mathbb{R} \quad \text{and} \quad \frac{\alpha}{\beta} > 0.$$

THEOREM 2.1.1. $\quad \triangle z_1 z_2 z_3 \sim \triangle w_1 w_2 w_3$

$$\Longleftrightarrow \frac{z_2 - z_1}{z_3 - z_1} = \frac{w_2 - w_1}{w_3 - w_1}$$

$$\Longleftrightarrow \begin{vmatrix} z_1 & w_1 & 1 \\ z_2 & w_2 & 1 \\ z_3 & w_3 & 1 \end{vmatrix} = 0.$$

Proof. Two triangles are similar if and only if the ratios of the lengths of the two corresponding sides are the same and the (corresponding) angles between them are the same (including the orientation). Hence

$$\triangle z_1 z_2 z_3 \sim \triangle w_1 w_2 w_3$$

$$\Longleftrightarrow \left| \frac{z_2 - z_1}{z_3 - z_1} \right| = \left| \frac{w_2 - w_1}{w_3 - w_1} \right| \quad \text{and} \quad \arg \frac{z_2 - z_1}{z_3 - z_1} = \arg \frac{w_2 - w_1}{w_3 - w_1}$$

$$\Longleftrightarrow \frac{z_2 - z_1}{z_3 - z_1} = \frac{w_2 - w_1}{w_3 - w_1}$$

$$\Longleftrightarrow \begin{vmatrix} z_1 & w_1 & 1 \\ z_2 & w_2 & 1 \\ z_3 & w_3 & 1 \end{vmatrix} = 0.$$

$\square$

COROLLARY 2.1.2. $\triangle z_1 z_2 z_3 \sim \triangle w_1 w_2 w_3$ *(reversed)*

$$\Longleftrightarrow \frac{z_2 - z_1}{z_3 - z_1} = \frac{\overline{w}_2 - \overline{w}_1}{\overline{w}_3 - \overline{w}_1}$$

$$\Longleftrightarrow \begin{vmatrix} z_1 & \overline{w}_1 & 1 \\ z_2 & \overline{w}_2 & 1 \\ z_3 & \overline{w}_3 & 1 \end{vmatrix} = 0.$$

Proof. $\because \triangle \overline{w}_1 \overline{w}_2 \overline{w}_3 \sim \triangle w_1 w_2 w_3$ (reversed)

$\therefore \triangle z_1 z_2 z_3 \sim \triangle w_1 w_2 w_3$ (reversed) $\Longleftrightarrow \triangle z_1 z_2 z_3 \sim \triangle \overline{w}_1 \overline{w}_2 \overline{w}_3.$ $\square$

EXAMPLE. Three points z_1, z_2, z_3 are collinear

$$\Longleftrightarrow \triangle z_1 z_2 z_3 \sim \triangle \overline{z}_1 \overline{z}_2 \overline{z}_3$$

$$\Longleftrightarrow \begin{vmatrix} z_1 & \overline{z}_1 & 1 \\ z_2 & \overline{z}_2 & 1 \\ z_3 & \overline{z}_3 & 1 \end{vmatrix} = 0.$$

EXAMPLE. Given two distinct points z_1 and z_2, find the equations of
(a) the line passing through the points z_1 and z_2 ;
(b) the perpendicular bisector of the line segment joining z_1 and z_2.

Solutions.

(a)
$$\begin{vmatrix} z & \overline{z} & 1 \\ z_1 & \overline{z}_1 & 1 \\ z_2 & \overline{z}_2 & 1 \end{vmatrix} = 0.$$

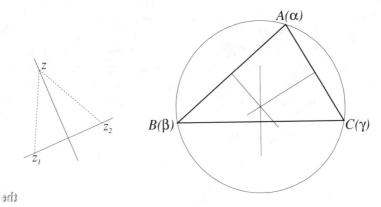

FIGURE 2.2

(b)
$$\begin{vmatrix} z & \bar{z} & 1 \\ z_1 & \bar{z}_2 & 1 \\ z_2 & \bar{z}_1 & 1 \end{vmatrix} = 0.$$

EXAMPLE. The perpendicular bisectors of the three sides of an arbitrary triangle meet at a point. This point is called the *circumcenter* of the triangle.

Solution. Let the three vertices A, B, C of the triangle be represented by complex numbers α, β, γ, respectively. Then the equation of the perpendicular bisector of the side BC is

$$\begin{vmatrix} z & \bar{z} & 1 \\ \beta & \bar{\gamma} & 1 \\ \gamma & \bar{\beta} & 1 \end{vmatrix} = 0;$$

i.e.,

$$(\bar{\beta} - \bar{\gamma})z + (\beta - \gamma)\bar{z} = |\beta|^2 - |\gamma|^2.$$

Similarly, those of the sides CA and AB are

$$(\bar{\gamma} - \bar{\alpha})z + (\gamma - \alpha)\bar{z} = |\gamma|^2 - |\alpha|^2,$$

$$(\bar{\alpha} - \bar{\beta})z + (\alpha - \beta)\bar{z} = |\alpha|^2 - |\beta|^2,$$

respectively. Adding any two of these three equations gives the third one, which implies that the solution of any two of these equations automatically satisfies the third. In other words, the intersection of any two perpendicular bisectors is on the remaining perpendicular bisector.

Solving a system of simultaneous equations consisting of any two of these three equations, we obtain the circumcenter :

$$z = \frac{|\alpha|^2(\beta - \gamma) + |\beta|^2(\gamma - \alpha) + |\gamma|^2(\alpha - \beta)}{\overline{\alpha}(\beta - \gamma) + \overline{\beta}(\gamma - \alpha) + \overline{\gamma}(\alpha - \beta)}.$$

Note that, by symmetry, we see again that this solution also satisfies the remaining equation.

Example. $\triangle z_1 z_2 z_3$ is an equilateral triangle

$$\Longleftrightarrow \triangle z_1 z_2 z_3 \sim \triangle z_3 z_1 z_2$$

$$\Longleftrightarrow \begin{vmatrix} z_1 & z_3 & 1 \\ z_2 & z_1 & 1 \\ z_3 & z_2 & 1 \end{vmatrix} = 0$$

$$\Longleftrightarrow z_1^2 + z_2^2 + z_3^2 - z_2 z_3 - z_3 z_1 - z_1 z_2 = 0$$

$$\Longleftrightarrow (z_1 + \omega z_2 + \omega^2 z_3) \cdot (z_1 + \omega^2 z_2 + \omega z_3) = 0 \qquad (\omega^2 + \omega + 1 = 0)$$

$$\Longleftrightarrow z_1 + \omega z_2 + \omega^2 z_3 = 0 \qquad \text{or} \qquad z_1 + \omega^2 z_2 + \omega z_3 = 0$$

$$\Longleftrightarrow \begin{vmatrix} z_1 & 1 & 1 \\ z_2 & \omega & 1 \\ z_3 & \omega^2 & 1 \end{vmatrix} = 0 \quad \text{or} \quad \begin{vmatrix} z_1 & 1 & 1 \\ z_2 & \omega^2 & 1 \\ z_3 & \omega & 1 \end{vmatrix} = 0.$$

$$\Longleftrightarrow \triangle z_1 z_2 z_3 \sim \triangle 1 \omega \omega^2 \qquad \text{or} \qquad \triangle z_1 z_2 z_3 \sim \triangle 1 \omega^2 \omega.$$

Example.   (Napoleon) On each side of an arbitrary triangle, draw an exterior equilateral triangle. Then the centroids of these three equilateral triangles are the vertices of a fourth equilateral triangle.

Proof. Let $\triangle z_1 z_2 z_3$ be the given triangle, and

$$\triangle w_1 z_3 z_2, \qquad \triangle z_3 w_2 z_1, \qquad \triangle z_2 z_1 w_3$$

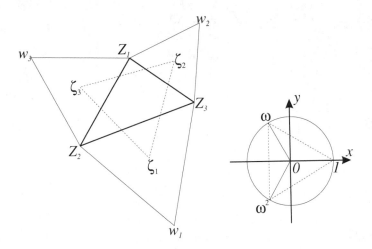

FIGURE 2.3

be all equilateral with the same orientation as $\triangle 1\omega\omega^2$, say, (where $\omega^2 + \omega + 1 = 0$), and with ζ_1, ζ_2, ζ_3 as the centroids of these equilateral triangles. Then

$$w_1 + \omega z_3 + \omega^2 z_2 = 0,$$

$$z_3 + \omega w_2 + \omega^2 z_1 = 0,$$

$$z_2 + \omega z_1 + \omega^2 w_3 = 0.$$

To prove that $\triangle \zeta_1\zeta_2\zeta_3$ is equilateral, we compute

$\zeta_1 + \omega\zeta_2 + \omega^2\zeta_3$

$$= \frac{1}{3}(w_1 + z_3 + z_2) + \frac{\omega}{3}(z_3 + w_2 + z_1) + \frac{\omega^2}{3}(z_2 + z_1 + w_3)$$

$$= \frac{1}{3}\left\{(w_1 + \omega z_3 + \omega^2 z_2) + (z_3 + \omega w_2 + \omega^2 z_1) + (z_2 + \omega z_1 + \omega^2 w_3)\right\}$$

$$= 0.$$

Therefore, $\triangle \zeta_1\zeta_2\zeta_3$ is an equilateral triangle. $\square$

Alternate Proof. Since $\triangle \zeta_1 z_3 z_2 \sim \triangle 01\omega$, we have

$$\begin{vmatrix} \zeta_1 & 0 & 1 \\ z_3 & 1 & 1 \\ z_2 & \omega & 1 \end{vmatrix} = 0;$$

i.e.,

$$(1 - \omega)\zeta_1 - z_2 + \omega z_3 = 0.$$

$$\therefore \zeta_1 = \frac{z_2 - \omega z_3}{1 - \omega}.$$

Similarly,

$$\zeta_2 = \frac{z_3 - \omega z_1}{1 - \omega}, \qquad \zeta_3 = \frac{z_1 - \omega z_2}{1 - \omega}.$$

$$\therefore \zeta_1 + \omega \zeta_2 + \omega^2 \zeta_3 = \frac{1}{1 - \omega} \left\{ (z_2 - \omega z_3) + \omega(z_3 - \omega z_1) + \omega^2(z_1 - \omega z_2) \right\}$$

$$= 0.$$

$\square$

The above result is usually attributed to Napoleon. It is well known that Napoleon established l'École polytechnique (1794), which produced most of the French mathematicians in the early 19th century, and was fond of mathematics, especially geometry. Yet many people are skeptical that Napoleon knew enough geometry to discover this theorem. Incidentally, Napoleon Bonaparte is one of very few people in modern history who are known by their first names. Galileo Galilei (1564–1642) is another example.

EXAMPLE.   Lines ℓ_1, ℓ_2, ℓ_3 are parallel to each other with ℓ_2 between ℓ_1 and ℓ_3. The distance between ℓ_1 and ℓ_2 is a, and that between ℓ_2 and ℓ_3 is b. Express the area of an equilateral triangle having one vertex on each of the three parallel lines in terms of a and b.

Solution.   Choose the coordinates as in Figure 2.4. We fix one vertex at ai, move another vertex t along the real axis, and try to find the locus

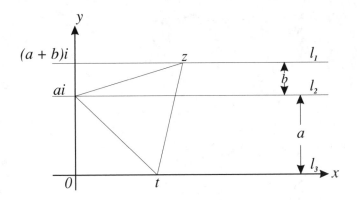

FIGURE 2.4

of the third vertex z. We have

$$z + ai\omega + t\omega^2 = 0 \qquad (\omega^2 + \omega + 1 = 0).$$

$$\therefore z = \frac{1}{2}\left\{(a\sqrt{3} + t) + i(a + t\sqrt{3})\right\}.$$

When the third vertex z is on the line $\Im z = a + b$, we have

$$\frac{1}{2}(a + t\sqrt{3}) = a + b; \qquad \text{i.e.,} \qquad t = \frac{1}{\sqrt{3}}(a + 2b).$$

Therefore, the square of the length of a side is

$$|t - ai|^2 = \frac{1}{3}(a + 2b)^2 + a^2 = \frac{4}{3}(a^2 + ab + b^2).$$

Hence the desired area is

$$\frac{1}{2} \cdot \frac{4}{3}(a^2 + ab + b^2) \cdot \sin\frac{\pi}{3} = \frac{\sqrt{3}}{3}(a^2 + ab + b^2).$$

2.2 The Ptolemy–Euler Theorem

For any four complex numbers $\alpha, \beta, \gamma, \delta$, the following identity is easy to verify:

$$(\alpha - \beta) \cdot (\gamma - \delta) + (\alpha - \delta) \cdot (\beta - \gamma) = (\alpha - \gamma) \cdot (\beta - \delta).$$

By the triangle inequality, we obtain

$$|\alpha - \beta| \cdot |\gamma - \delta| + |\alpha - \delta| \cdot |\beta - \gamma| \geq |\alpha - \gamma| \cdot |\beta - \delta|.$$

Let us investigate when the inequality becomes an equality. In the case of the triangle inequality,

$$|z_1 + z_2| \leq |z_1| + |z_2|,$$

equality holds if and only if $\frac{z_1}{z_2}$ is a positive real number (provided $z_1 z_2 \neq 0$). Thus we are looking for a condition to ensure that $\frac{(\alpha - \beta)(\gamma - \delta)}{(\alpha - \delta)(\beta - \gamma)}$ is a positive real number. But

$$\frac{(\alpha - \beta)(\gamma - \delta)}{(\alpha - \delta)(\beta - \gamma)} \quad \text{is a positive real number}$$

$$\Longleftrightarrow \frac{\alpha - \beta}{\alpha - \delta} \bigg/ \frac{\gamma - \beta}{\gamma - \delta} \quad \text{is a negative real number}$$

$$\Longleftrightarrow \arg\left\{ \frac{\alpha - \beta}{\alpha - \delta} \bigg/ \frac{\gamma - \beta}{\gamma - \delta} \right\}$$

$$= \arg\left\{ \frac{\alpha - \beta}{\alpha - \delta} \right\} - \arg\left\{ \frac{\gamma - \beta}{\gamma - \delta} \right\} \equiv \pi \pmod{2\pi}.$$

That is, $\alpha, \beta, \gamma, \delta$ are cocyclic (see Corollary A.2.3 in Appendix A) *and* α and γ are on the opposite sides of the chord joining β and δ, which results in the alphabetical order (clockwise or counterclockwise).

We have proven the following

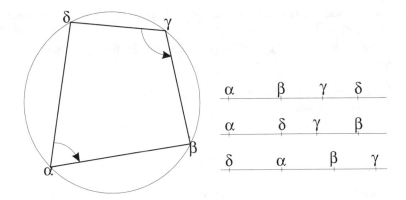

FIGURE 2.5

THEOREM 2.2.1. *For any four points A, B, C, D in the plane,*

$$\overline{AB}\cdot\overline{CD} + \overline{BC}\cdot\overline{DA} \geq \overline{AC}\cdot\overline{BD}.$$

Equality holds if and only if these four points are cocyclic (or collinear) and are in alphabetical order (clockwise or counterclockwise).

The equality was discovered by C. Ptolemy (ca. 85–165), while the general case was found over a thousand years later by L. Euler (1707–1783). However, using complex numbers, their results can be obtained in a single stroke.

The expression

$$(\alpha, \beta;\, \gamma, \delta) := \left(\frac{\alpha - \gamma}{\alpha - \delta}\right) \Big/ \left(\frac{\beta - \gamma}{\beta - \delta}\right)$$

is called the *cross ratio* of the four points α, β, γ, δ. It plays an important role in various parts of mathematics, especially in projective geometry, which is certainly one of the most beautiful branches of mathematics.

COROLLARY 2.2.2. *Four points α, β, γ, δ, are cocyclic (or collinear) if and only if*

$$(\alpha, \beta;\, \gamma, \delta) \in \mathbb{R}.$$

In the sequel, *'collinear' is regarded as a particular (degenerate) case of 'cocyclic'.*

When the inscribing quadrangle is a rectangle, the Ptolemy theorem reduces to

COROLLARY 2.2.3 (Pythagoras). *In a right triangle ABC, with the angle at C being the right angle,*

$$\overline{BC}^2 + \overline{CA}^2 = \overline{AB}^2.$$

EXAMPLE. Let $ABCDE$ be a regular pentagon of sides ℓ inscribed in a circle of radius r, and P be the midpoint of $\overset{\frown}{CD}$, and d the length of a diagonal. Applying the Ptolemy theorem 2.2.1 to the quadrangle $ACDE$ and $ACPD$, we get

$$d\ell + \ell^2 = d^2 \qquad \text{and} \qquad 2xd = 2r\ell,$$

where x is the length of a side of a regular decagon inscribed in the circle of radius r. It follows that

$$\varphi := \frac{r}{x} = \frac{d}{\ell} \qquad \text{satisfies} \qquad \varphi^2 = \varphi + 1.$$

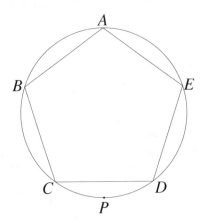

FIGURE 2.6

Therefore, the ratio of the radius r to the side x of the inscribed regular decagon is the famous *golden ratio* :

$$\varphi = \frac{1 + \sqrt{5}}{2} \qquad (\because \ \varphi > 0).$$

In particular, a regular pentagon and a regular decagon can be constructed with a compass and straightedge, as we have already seen at the end of §1.9.

2.3 The Clifford Theorems

In this section,[1] we prove an infinite sequence of theorems discovered by W. K. Clifford (1845–1879). The crucial step is the following lemma, which we shall use in other sections too.

LEMMA 2.3.1. *Suppose there are four circles* C_1, C_2, C_3, C_4 *in a plane. Let* C_1 *and* C_2 *intersect at* z_1 *and* w_1, C_2 *and* C_3 *intersect at* z_2 *and* w_2, C_3 *and* C_4 *intersect at* z_3 *and* w_3, C_4 *and* C_1 *intersect at* z_4 *and* w_4. *Then the points* z_1, z_2, z_3, z_4 *are cocyclic if and only if* w_1, w_2, w_3, w_4 *are cocyclic.*

Proof. By assumption, the following four cross ratios are real :

$$(z_1, w_2;\ z_2, w_1) = \frac{z_1 - z_2}{w_2 - z_2} \bigg/ \frac{z_1 - w_1}{w_2 - w_1},$$

$$(z_2, w_3;\ z_3, w_2) = \frac{z_2 - z_3}{w_3 - z_3} \bigg/ \frac{z_2 - w_2}{w_3 - w_2},$$

$$(z_3, w_4;\ z_4, w_3) = \frac{z_3 - z_4}{w_4 - z_4} \bigg/ \frac{z_3 - w_3}{w_4 - w_3},$$

$$(z_4, w_1;\ z_1, w_4) = \frac{z_4 - z_1}{w_1 - z_1} \bigg/ \frac{z_4 - w_4}{w_1 - w_4}.$$

---

[1] Other than the lemma, the material in this section will not be needed for the rest of the book, and so may be skipped in the first reading.

Therefore,

$$\frac{(z_1, w_2;\ z_2, w_1)}{(z_2, w_3;\ z_3, w_2)} \cdot \frac{(z_3, w_4;\ z_4, w_3)}{(z_4, w_1;\ z_1, w_4)}$$

$$= \left\{ \left(\frac{z_1 - z_2}{z_3 - z_2}\right) \Big/ \left(\frac{z_1 - z_4}{z_3 - z_4}\right) \right\} \cdot \left\{ \left(\frac{w_1 - w_2}{w_3 - w_2}\right) \Big/ \left(\frac{w_1 - w_4}{w_3 - w_4}\right) \right\}$$

$$= (z_1, z_3;\ z_2, z_4) \cdot (w_1, w_3;\ w_2, w_4)$$

is real. Hence $(z_1, z_3;\ z_2, z_4)$ is real if and only if $(w_1, w_3;\ w_2, w_4)$ is real.
□

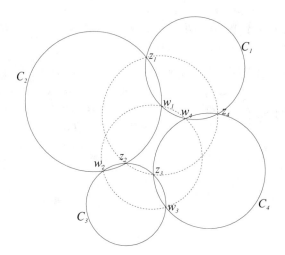

FIGURE 2.7

We say n lines in a plane are *in general position* if no two of them are parallel and no three of them meet at a point.

Let us call the intersection of two lines in general position, their *Clifford point*. From three lines in general position, we obtain three Clifford points, by choosing a pair of them each time, and the circle through these three points (the circumcircle of the triangle formed by these three lines) is called the *Clifford circle* of the three lines.

Now, given four lines C_1, C_2, C_3, C_4 in general position, let z_{jk} be the intersection of the lines C_j and C_k (other than ∞), and C_{lmn} be the circumcircle of $\triangle z_{mn} z_{nl} z_{lm}$. (We disregard the permutation of indices; for example, $z_{jk} = z_{kj}$, $C_{lmn} = C_{nlm}$.) Applying Lemma 2.3.1 to C_{234}, C_2, C_1, C_{134}, and noting that

C_{234},	C_2	intersect at	z_{23}, z_{24};
C_2,	C_1	intersect at	∞, z_{12};
C_1,	C_{134}	intersect at	z_{13}, z_{14};
C_{134},	C_{234}	intersect at	z_{34}, z_{1234},

where z_{1234} is the 'new' intersection of C_{234} and C_{134} (i.e., other than z_{34}), then, since z_{23}, ∞, z_{13}, z_{34} are collinear (all on C_3), we conclude that z_{24}, z_{12}, z_{14}, z_{1234} are cocyclic. But the circumcircle of $\triangle z_{24} z_{12} z_{14}$ is the circle C_{124}, hence the circles C_{234}, C_{134}, C_{124} meet at z_{1234}.

On the other hand, with the same C_{234}, C_2, C_1, C_{134}, if we note that z_{24}, ∞, z_{14}, z_{34} are collinear (all on C_4), we conclude that z_{23}, z_{12}, z_{13}, z_{1234} are cocyclic. But the circumcircle of $\triangle z_{23} z_{12} z_{13}$ is the circle C_{123}, hence the circles C_{234}, C_{134}, C_{123} meet at z_{1234}.

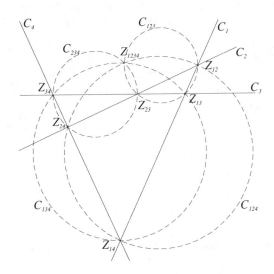

FIGURE 2.8

We have shown that the circles C_{234}, C_{134}, C_{124}, C_{123} meet at one point z_{1234}, which we call the *Clifford point* of the four lines C_1, C_2, C_3, C_4.

Before we proceed to the case of five lines, we remark that the point z_{jk} is at the intersection of the lines C_j and C_k ; the circle C_{lmn} passes through the points z_{mn}, z_{ln}, z_{lm} ; the point z_{klmn} is the intersection of the circles C_{lmn}, C_{kmn}, C_{kln}, C_{klm}. In particular, the circles C_{lmn} and C_{kmn} intersect at z_{klmn} and z_{mn}. Now we are ready for the next step.

Suppose we are given five lines C_1, C_2, C_3, C_4, C_5 in general position. Preserving our notations above, and taking four lines at a time, we obtain five Clifford points z_{2345}, z_{1345}, z_{1245}, z_{1235}, z_{1234}. We claim that these five Clifford points are cocyclic. To prove this, it is sufficient to prove that any four of these five Clifford points are cocyclic. For example, take z_{1345}, z_{1245}, z_{1235}, z_{1234}. These can be considered as the intersections of C_{134} and C_{135}, C_{125} and C_{145}, C_{123} and C_{125}, C_{124} and C_{134}, respectively. The second point of the intersection of these pairs of circles are the points z_{13}, z_{15}, z_{12}, z_{14}, which are collinear (all on C_1), and so, by Lemma 2.3.1, we get the desired result. The circle so obtained is called the *Clifford circle* of the lines C_1, C_2, C_3, C_4, C_5, and is denoted by C_{12345}.

Now given six lines C_1, C_2, C_3, C_4, C_5, C_6, in general position. Taking five lines at a time, we obtain six Clifford circles. We claim these six circles meet at a point, the *Clifford point* of the six lines. To prove this, it is sufficient to show that any three of these six circles meet at a point. Note carefully: it is simple to give an example of four circles, any three of which meet at a point without all four of them meeting at a point, but this cannot be done if we have five circles or more.

Suppose we want to show that C_{23456}, C_{13456}, C_{12456} meet at a point. Consider a sequence of four circles C_{23456}, C_{245}, C_{145}, C_{13456}. These intersect, in pairs, at z_{2345} and z_{2456}, z_{45} and z_{1245}, z_{1345} and z_{1456}, z_{3456} and z_{123456}, where z_{123456} is the intersection of C_{23456} and C_{13456} other than z_{3456}. But the points z_{2345}, z_{45}, z_{1345}, z_{3456} are all on the circle C_{345}. Hence the points z_{2456}, z_{1245}, z_{1456}, and z_{123456} must be cocyclic. However, the first three of these four points are on the circle C_{12456}, hence the circle C_{12456} passes through the intersection of C_{23456} and C_{13456}.

Now it is simple to carry on the induction argument.

2.4 The Nine-Point Circle

Given a triangle ABC, choose its circumcenter O to be the origin of the complex plane, and let α, β, γ be the complex numbers representing the vertices A, B, C, respectively. Without loss of generality, we may assume that the circumcircle has radius 1; i.e., $|\alpha| = |\beta| = |\gamma| = 1$. Then it is natural to ask, Where is the point $\sigma = \alpha + \beta + \gamma$?

Since $\sigma - \alpha = \beta + \gamma$, and $\dfrac{\beta + \gamma}{2}$ is the midpoint D of the side BC, σ is on the perpendicular from the vertex A to the side BC, and the length $|\sigma - \alpha|$ is twice that of OD. By symmetry, σ also is on the perpendicular from B to CA, and on that of C to AB; i.e., σ is the *orthocenter* H of $\triangle ABC$. Note that we have shown that three perpendiculars from the vertices to the opposite sides meet at a point, called the *orthocenter* of $\triangle ABC$.

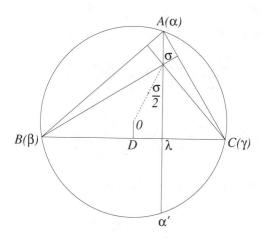

FIGURE 2.9

Now $\frac{\sigma}{2} = \frac{1}{2}(\alpha + \beta + \gamma)$ is the midpoint of the line segment joining the circumcenter O and the orthocenter H. The distance from $\frac{\sigma}{2}$ to the midpoint D of the side BC is

$$\left| \frac{\beta + \gamma}{2} - \frac{\sigma}{2} \right| = \left| \frac{\alpha}{2} \right| = \frac{1}{2}.$$

Similarly, the distance from $\frac{\sigma}{2}$ to the midpoint E of the side CA, and to the midpoint F of the side AB are all equal to $\frac{1}{2}$.

Furthermore, the distance from $\frac{\sigma}{2}$ to the midpoint of the line segment joining the orthocenter H to the vertex A is

$$\left| \frac{\alpha + \sigma}{2} - \frac{\sigma}{2} \right| = \frac{|\alpha|}{2} = \frac{1}{2}.$$

Similarly, the distance from $\frac{\sigma}{2}$ to the midpoint of BH, and to that of CH are also equal to $\frac{1}{2}$.

To find the foot λ of the perpendicular from the vertex A to the side BC, we first compute the point α' where this perpendicular meets the circumcircle again. Thus α' must satisfy the conditions

$$\overrightarrow{\alpha \alpha'} \perp \overrightarrow{\beta \gamma}, \qquad |\alpha'| = 1, \quad \alpha' \neq \alpha.$$

From the first condition, we get

$$\frac{\alpha - \alpha'}{\beta - \gamma} \qquad \text{is purely imaginary;}$$

i.e.,

$$\frac{\alpha - \alpha'}{\beta - \gamma} + \frac{\overline{\alpha} - \overline{\alpha'}}{\overline{\beta} - \overline{\gamma}} = 0.$$

Substituting the relations $\overline{\alpha} = \frac{1}{\alpha}$, etc., this becomes

$$\frac{\alpha - \alpha'}{\beta - \gamma} \left\{ 1 + \frac{\beta \gamma}{\alpha \alpha'} \right\} = 0.$$

Hence

$$\alpha' = -\frac{\beta \gamma}{\alpha}.$$

To check whether our computation is correct, note that $|\alpha'| = 1$, and $\frac{\beta}{\alpha} \cdot \frac{\gamma}{\alpha'} = -1$, and so $\arg \left(\frac{\beta}{\alpha} \right) + \arg \left(\frac{\gamma}{\alpha'} \right) = \pi$, which means that $\overrightarrow{\alpha \alpha'} \perp \overrightarrow{\beta \gamma}$; viz., α' is the point where the perpendicular from A to the side BC meets the circumcircle again.

Now the distances from the vertex B to α' and to σ are

$$\left| \beta + \frac{\beta\gamma}{\alpha} \right| = \left| \frac{\beta}{\alpha} \right| \cdot |\alpha + \gamma| = |\alpha + \gamma|,$$

$$|\sigma - \beta| = |(\alpha + \beta + \gamma) - \beta| = |\alpha + \gamma|,$$

respectively. Hence $\triangle\beta\alpha'\alpha$ is an isosceles triangle, and

$$\lambda = \frac{1}{2}(\sigma + \alpha') = \frac{1}{2}\left(\sigma - \frac{\beta\gamma}{\alpha} \right).$$

It follows that the distance from $\frac{\sigma}{2}$ to the foot λ (of the perpendicular from the vertex A to the side BC) is

$$\left| \lambda - \frac{\sigma}{2} \right| = \left| \frac{\alpha'}{2} \right| = \frac{1}{2}.$$

Similarly, the distances from $\frac{\sigma}{2}$ to the other two feet of perpendiculars are also $\frac{1}{2}$.

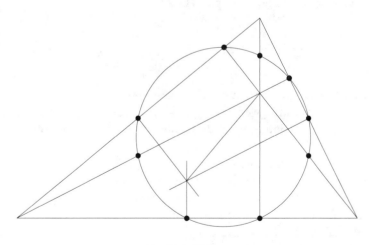

FIGURE 2.10

Summing up, we have obtained the following.

THEOREM 2.4.1 (The Nine-Point Circle). *In any triangle,*

(a) *the feet of the three perpendiculars from the vertices to the opposite sides;*

(b) *the midpoints of the three sides; and*

(c) *the midpoints of the segments joining the orthocenter to the three vertices,*

are all on the same circle, whose center is at the midpoint of the segment joining the orthocenter and the circumcenter, and the radius is one half of that of the circumcircle.

The line passing through the orthocenter, circumcenter, centroid, and the center of the nine-point circle is known as the *Euler line* of the triangle.

Let z_1, z_2, z_3 be three arbitrary points on the unit circle $|z| = 1$. Then the circumcenter, centroid, the center of the nine-point circle, and the orthocenter of $\triangle z_1 z_2 z_3$ are given by

$$0, \quad \frac{1}{3}(z_1 + z_2 + z_3), \quad \frac{1}{2}(z_1 + z_2 + z_3), \quad (z_1 + z_2 + z_3),$$

respectively, and the radius of the nine-point circle is $\frac{1}{2}$.

Suppose we are given four points z_1, z_2, z_3, z_4 on the unit circle. Choosing three points out of these four points at a time, we obtain four triangles (all of which are inscribed in the unit circle).

The center of the nine-point circle of $\triangle z_2 z_3 z_4$ is $\tau_1 = \frac{1}{2}(z_2 + z_3 + z_4)$, the center of the nine-point circle of $\triangle z_1 z_3 z_4$ is $\tau_2 = \frac{1}{2}(z_1 + z_3 + z_4)$, the center of the nine-point circle of $\triangle z_1 z_2 z_4$ is $\tau_3 = \frac{1}{2}(z_1 + z_2 + z_4)$, the center of the nine-point circle of $\triangle z_1 z_2 z_3$ is $\tau_4 = \frac{1}{2}(z_1 + z_2 + z_3)$, and their radii are all equal to $\frac{1}{2}$.

Consider the point

$$\tau = \frac{1}{2}(z_1 + z_2 + z_3 + z_4).$$

Then it is immediate that

$$|\tau - \tau_1| = |\tau - \tau_2| = |\tau - \tau_3| = |\tau - \tau_4| = \frac{1}{2}.$$

Hence the nine-point circles of

$$\triangle z_2 z_3 z_4, \qquad \triangle z_1 z_3 z_4, \qquad \triangle z_1 z_2 z_4, \qquad \triangle z_1 z_2 z_3,$$

all pass through the point

$$\tau = \frac{1}{2}(z_1 + z_2 + z_3 + z_4);$$

in particular, the centers of the four nine-point circles are on the circle with center at τ and radius $\frac{1}{2}$. Let us call this circle the *nine-point circle of the quadrangle* $z_1 z_2 z_3 z_4$.

Now, suppose we are given five points z_1, z_2, z_3, z_4, z_5 on the unit circle. Then the center of the nine-point circle of the quadrangle $z_2 z_3 z_4 z_5$ is

$$\mu_1 = \frac{1}{2}(z_2 + z_3 + z_4 + z_5), \qquad \text{etc.,}$$

and the distances from these centers to the point

$$\mu = \frac{1}{2}(z_1 + z_2 + z_3 + z_4 + z_5)$$

are

$$|\mu - \mu_1| = \frac{1}{2}, \qquad \text{etc.}$$

Hence the centers of the nine-point circles of the quadrangles

$$z_2 z_3 z_4 z_5, \qquad z_1 z_3 z_4 z_5, \qquad z_1 z_2 z_4 z_5, \qquad z_1 z_2 z_3 z_5, \qquad z_1 z_2 z_3 z_4$$

are on the circle with the center at $\mu = \frac{1}{2}(z_1 + z_2 + z_3 + z_4 + z_5)$ and radius $\frac{1}{2}$. Let us call this circle the *nine-point circle of the pentagon* $z_1 z_2 z_3 z_4 z_5$.

Next, suppose we are given six points $z_1, z_2, z_3, z_4, z_5, z_6$ on the unit circle, ...

Thus we have an infinite sequence of theorems discovered by J. L. Coolidge.

2.5   The Simson Line

We start this section with some preparatory comments about the equation of a line.

Given a line ℓ, let α be the unit vector perpendicular to ℓ, and p the distance from the origin to the line ℓ. Then for any point z on ℓ, $z - p\alpha$ is a vector on ℓ, and since α is a vector perpendicular to ℓ, we have

$$\frac{z - p\alpha}{\alpha} \quad \text{is purely imaginary; i.e.,}$$

$$\frac{z - p\alpha}{\alpha} + \frac{\bar{z} - p\bar{\alpha}}{\bar{\alpha}} = 0.$$

Hence the equation of the line ℓ is given by

$$\frac{z}{\alpha} + \frac{\bar{z}}{\bar{\alpha}} = 2p; \quad \text{i.e.,} \quad z + k\bar{z} = 2p\alpha,$$

where $p \in \mathbb{R}$, $k = \frac{\alpha}{\bar{\alpha}}$, $\therefore$ $|k| = 1$. And to obtain the equation of a line perpendicular to ℓ, simply replace α by $i\alpha$:

$$\frac{z}{i\alpha} - \frac{\bar{z}}{\overline{i\alpha}} = 2q \quad \text{for some} \quad q \in \mathbb{R}; \text{ i.e.,}$$

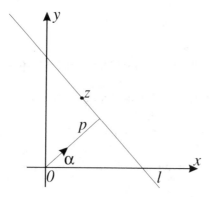

FIGURE 2.11

$$\frac{z}{\alpha} - \frac{\overline{z}}{\overline{\alpha}} = 2qi; \qquad \text{viz.,} \quad z - k\overline{z} = 2qi\alpha, \quad \left(k = \frac{\alpha}{\overline{\alpha}}\right),$$

where the constant on the right can be adjusted to pass through a specific point.

Note that a linear equation in z and $\overline{z}$ jointly is the equation of a line if and only if it is *self-conjugate* ; viz., the relation obtained by taking the complex conjugate of both sides of the equation must be equivalent to the original equation. For example, taking the complex conjugate of

$$z + k\overline{z} = 2p\alpha \qquad \left(\text{where} \quad p \in \mathbb{R}, \quad k = \frac{\alpha}{\overline{\alpha}}\right),$$

we obtain

$$\overline{z} + \overline{k}z = 2p\overline{\alpha}.$$

Substituting $\overline{k} = \frac{1}{k}$, this becomes

$$k\overline{z} + z = 2p\overline{\alpha}k = 2p\alpha,$$

which is the original equation. Similarly, for the equation

$$z - k\overline{z} = 2qi\alpha \qquad \left(q \in \mathbb{R}, \quad k = \frac{\alpha}{\overline{\alpha}}\right).$$

In particular, it is necessary (but not sufficient) that the coefficients α and β of z and $\overline{z}$ in

$$\alpha z + \beta\overline{z} = \gamma$$

have the same absolute value; i.e., $|\alpha| = |\beta|$. It follows that equations such as $z + \overline{z} = i$ or $2z - \overline{z} = 1$ are not equations of lines.

We are now ready to prove the Simson theorem.

THEOREM 2.5.1. *Given $\triangle ABC$ and a point D, let P, Q, R be the feet of the perpendiculars from the point D to the sides BC, CA, AB, respectively. Then the points P, Q, R are collinear if and only if D is on the circumcircle of $\triangle ABC$.*

Proof. Without loss of generality, we may assume that $\triangle ABC$ is inscribed in the unit circle, and the points A, B, C, D are represented by

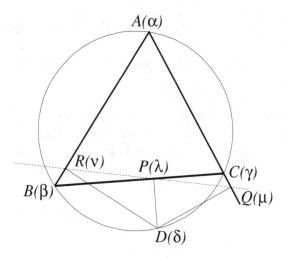

FIGURE 2.12

the complex numbers α, β, γ, δ, respectively. Then the equation of the line BC is

$$\begin{vmatrix} z & \bar{z} & 1 \\ \beta & \bar{\beta} & 1 \\ \gamma & \bar{\gamma} & 1 \end{vmatrix} = 0;$$

i.e.,

$$(\bar{\beta} - \bar{\gamma})z - (\beta - \gamma)\bar{z} + (\beta\bar{\gamma} - \bar{\beta}\gamma) = 0.$$

Using the relations $\bar{\beta} = \frac{1}{\beta}, \bar{\gamma} = \frac{1}{\gamma}$, this can be rewritten as

$$z + \beta\gamma\bar{z} = \beta + \gamma.$$

Hence the equation of the perpendicular from $D(\delta)$ to the side BC is

$$z - \beta\gamma\bar{z} = \delta - \beta\gamma\bar{\delta}.$$

Therefore, the intersection $P(\lambda)$ of these two lines is obtained by solving these two equations :

$$\lambda = \frac{1}{2}(\beta + \gamma + \delta - \beta\gamma\bar{\delta}).$$

Similarly, $Q(\mu)$, $R(\nu)$ are given by

$$\mu = \frac{1}{2}(\gamma + \alpha + \delta - \gamma\alpha\overline{\delta}),$$

$$\nu = \frac{1}{2}(\alpha + \beta + \delta - \alpha\beta\overline{\delta}).$$

Now,

$$P(\lambda), Q(\mu), R(\nu) \text{ are collinear} \iff \frac{\lambda - \nu}{\mu - \nu} \in \mathbb{R}.$$

However, with the notation $r = |\delta|$ (hence $\overline{\delta} = \frac{r^2}{\delta}$), we have

$$\frac{\lambda - \nu}{\mu - \nu} = \frac{(\gamma - \alpha)(1 - \beta\overline{\delta})}{(\gamma - \beta)(1 - \alpha\overline{\delta})}$$

$$= \left(\frac{\alpha - \gamma}{\beta - \gamma}\right) \Bigg/ \left(\frac{\alpha - \delta r^{-2}}{\beta - \delta r^{-2}}\right)$$

$$= (\alpha, \beta; \gamma, \delta r^{-2}).$$

Therefore,

$$P, Q, R \text{ are collinear} \iff (\alpha, \beta; \gamma, \delta r^{-2}) \in \mathbb{R}$$

$$\iff \alpha, \beta, \gamma, \delta r^{-2} \text{ are cocyclic}$$

$$\iff |\delta r^{-2}| = 1$$

$$\iff r = |\delta| = 1.$$

$\square$

This line is usually called the *Simson line* of the point D with respect to $\triangle ABC$. However, historians have searched in vain for it through the works of Robert Simson (1687–1768). It seems to have been published first by William Wallace (1768–1843) in 1797.

We now try to find the equation of the Simson line. We keep the same notations as before; in particular, we assume $\triangle ABC$ is inscribed in the unit circle, and the point $D(\delta)$ is on the unit circle. Then the foot P of the perpendicular from $D(\delta)$ to the side BC is given by

$$z = \frac{1}{2}\left(\beta + \gamma + \delta - \frac{\beta\gamma}{\delta}\right).$$

Let us now introduce the notations

$$\sigma_1 = \alpha + \beta + \gamma, \qquad \sigma_2 = \beta\gamma + \gamma\alpha + \alpha\beta, \qquad \sigma_3 = \alpha\beta\gamma;$$

then

$$\overline{\sigma}_1 = \overline{\alpha} + \overline{\beta} + \overline{\gamma} = \frac{1}{\alpha} + \frac{1}{\beta} + \frac{1}{\gamma} = \frac{\sigma_2}{\sigma_3},$$

$$\overline{\sigma}_3 = \overline{\alpha}\overline{\beta}\overline{\gamma} = \frac{1}{\alpha\beta\gamma} = \frac{1}{\sigma_3}.$$

Therefore, the above expression for z becomes

$$z = \frac{1}{2}\left(\sigma_1 - \alpha + \delta - \frac{\sigma_3}{\delta\alpha}\right),$$

and

$$\overline{z} = \frac{1}{2}\left(\overline{\sigma}_1 - \overline{\alpha} + \overline{\delta} - \frac{\overline{\sigma}_3}{\overline{\delta}\overline{\alpha}}\right)$$

$$= \frac{1}{2}\left(\frac{\sigma_2}{\sigma_3} - \frac{1}{\alpha} + \frac{1}{\delta} - \frac{\delta\alpha}{\sigma_3}\right).$$

Eliminating α from these two relations, we get

$$\delta z - \sigma_3 \overline{z} = \frac{1}{2}\left(\delta^2 + \sigma_1\delta - \sigma_2 - \frac{\sigma_3}{\delta}\right).$$

This is a relation that must be satisfied by the foot $P(\lambda)$ of the perpendicular from $D(\delta)$ to the side BC. However, since this relation contains σ_1, σ_2, σ_3 only, and so is symmetric with respect to α, β, γ. It follows that this relation is also satisfied by the feet $Q(\mu)$ and $R(\nu)$ of the perpendiculars from $D(\delta)$ to the sides CA and AB, respectively.

However, this is an equation of a straight line, hence the feet P, Q, R are collinear, and the equation we obtained is the equation of the Simson line. We have given an alternate proof to the *if*-part of Theorem 2.5.1.

THEOREM 2.5.2. *Let* L, M, N *be three points on the circumcircle of* $\triangle ABC$. *The necessary and sufficient condition that the Simson lines of the points* L, M, N *with respect to* $\triangle ABC$ *meet at one point is*

$$\overset{\frown}{AL} + \overset{\frown}{BM} + \overset{\frown}{CN} \equiv 0 \quad (\text{mod } 2\pi).$$

Proof. Let the circumcircle of $\triangle ABC$ be the unit circle, and u_1, u_2, u_3 the complex numbers representing the points L, M, N, respectively. Then the equations of the three Simson lines under consideration are

$$u_1 z - \sigma_3 \bar{z} = \frac{1}{2} \left(u_1^2 + \sigma_1 u_1 - \sigma_2 - \frac{\sigma_3}{u_1} \right),$$

$$u_2 z - \sigma_3 \bar{z} = \frac{1}{2} \left(u_2^2 + \sigma_1 u_2 - \sigma_2 - \frac{\sigma_3}{u_2} \right),$$

$$u_3 z - \sigma_3 \bar{z} = \frac{1}{2} \left(u_3^2 + \sigma_1 u_3 - \sigma_2 - \frac{\sigma_3}{u_3} \right).$$

Hence the intersection of the first two Simson lines is given by

$$z = \frac{1}{2} \left(u_1 + u_2 + \sigma_1 + \frac{\sigma_3}{u_1 u_2} \right),$$

and that of the last two Simson lines is given by

$$z = \frac{1}{2} \left(u_2 + u_3 + \sigma_1 + \frac{\sigma_3}{u_2 u_3} \right).$$

Therefore, the necessary and sufficient condition for these two points to coincide is that $\sigma_3 = u_1 u_2 u_3$; i.e., $\alpha\beta\gamma = u_1 u_2 u_3$. Since α, β, γ, u_1, u_2, u_3 are all complex numbers with absolute value 1, by setting their arguments as θ_1, θ_2, θ_3, φ_1, φ_2, φ_3, respectively, we obtain

$$\theta_1 + \theta_2 + \theta_3 \equiv \varphi_1 + \varphi_2 + \varphi_3 \quad (\text{mod } 2\pi).$$

$$\therefore \quad (\theta_1 - \varphi_1) + (\theta_2 - \varphi_2) + (\theta_3 - \varphi_3) \equiv 0 \quad (\bmod\ 2\pi),$$

which is the desired condition.   □

Note that if this condition is satisfied, then the intersection is given by

$$z = \frac{1}{2}(\sigma_1 + u_1 + u_2 + u_3)$$

$$= \frac{1}{2}(\alpha + \beta + \gamma + u_1 + u_2 + u_3).$$

By symmetry, we obtain the following

COROLLARY 2.5.3. *Let A, B, C, L, M, N be six points on a circle. Then the Simson lines of the points L, M, N with respect to $\triangle ABC$ meet at a point if and only if the Simson lines of A, B, C with respect to $\triangle LMN$ meet at a point. Moreover, in this case, all six Simson lines meet at the midpoint of the line segment joining the orthocenters of $\triangle ABC$ and $\triangle LMN$.*

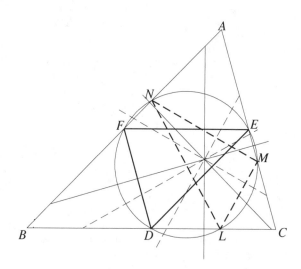

FIGURE 2.13

COROLLARY 2.5.4. *Let D, E, F be the midpoints of the respective sides BC, CA, AB of $\triangle ABC$, and L, M, N the feet of the perpendiculars from the vertices A, B, C to the opposite sides, respectively. Then the six points D, E, F, L, M, N are on the nine-point circle of $\triangle ABC$, and the Simson lines of the points L, M, N with respect to $\triangle DEF$ meet at a point. The converse is also true.*

Proof. Let the circumcircle of $\triangle ABC$ be the unit circle, and the vertices A, B, C be represented by the complex numbers α, β, γ, respectively. Then from our discussion of the nine-point circle, the points L, M, N are given by

$$\frac{1}{2}\left(\sigma_1 - \frac{\beta\gamma}{\alpha}\right), \quad \frac{1}{2}\left(\sigma_1 - \frac{\gamma\alpha}{\beta}\right), \quad \frac{1}{2}\left(\sigma_1 - \frac{\alpha\beta}{\gamma}\right);$$

and the points D, E, F are given by

$$\frac{1}{2}(\beta + \gamma), \quad \frac{1}{2}(\gamma + \alpha), \quad \frac{1}{2}(\alpha + \beta).$$

Moreover, all six points are on the nine-point circle whose center K is at $\frac{\sigma_1}{2}$, and the radius is $\frac{1}{2}$. Therefore, we have

$$\overrightarrow{KL}: -\frac{\beta\gamma}{2\alpha}, \quad \overrightarrow{KM}: -\frac{\gamma\alpha}{2\beta}, \quad \overrightarrow{KN}: -\frac{\alpha\beta}{2\gamma},$$

$$\overrightarrow{KD}: -\frac{\alpha}{2}, \quad \overrightarrow{KE}: -\frac{\beta}{2}, \quad \overrightarrow{KF}: -\frac{\gamma}{2}.$$

Since

$$\left(-\frac{\beta\gamma}{2\alpha}\right) \cdot \left(-\frac{\gamma\alpha}{2\beta}\right) \cdot \left(-\frac{\alpha\beta}{2\gamma}\right) = \left(-\frac{\alpha}{2}\right) \cdot \left(-\frac{\beta}{2}\right) \cdot \left(-\frac{\gamma}{2}\right),$$

the condition in the theorem is satisfied. $\square$

2.6 Generalizations of the Simson Theorem

We now present another proof of the Simson theorem 2.5.1 that gives a generalization at no extra cost.

THEOREM 2.6.1. *Let P, Q, R be the feet of the perpendiculars from an arbitrary point D to the respective sides BC, CA, AB of $\triangle ABC$. Then P, Q, R are collinear if and only if the point D is on the circumcircle of $\triangle ABC$.*

Proof. Since $\angle DPC = \angle DRB (= \frac{\pi}{2})$, the points P, R, B, D are cocyclic. Call this circle S_B. Similarly, the points P, Q, C, D are also cocyclic. Call this circle S_C. We now apply Lemma 2.3.1 for the Clifford theorems to S_B, AB, AC, S_C:

The circle S_B and the line AB intersect at B and R;
the lines AB and AC intersect at A and ∞;
the line AC and the circle S_C intersect at C and Q;
the circles S_C and S_B intersect at D and P.

Therefore,

$$B, A, C, D \quad \text{are cocyclic} \iff R, \infty, Q, P \quad \text{are cocyclic}$$

$$\iff P, Q, R \quad \text{are collinear}$$

$\square$

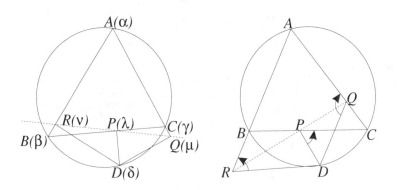

FIGURE 2.14

Theorem 2.6.2. *Let P, Q, R be points on (the extensions of) the respective sides BC, CA, AB of △ABC. Suppose D is a point with the property that*

$$\angle DPC \equiv \angle DQA \equiv \angle DRB \pmod{\pi},$$

*where all angles are considered **oriented**. Then the points P, Q, R are collinear if and only if the point D is on the circumcircle of △ABC.*

Proof. Same as the previous theorem. □

As a corollary, we obtain the following.

Theorem 2.6.3 (Aubert). *Let A, A', B, B', C, C', D be seven cocyclic points such that AA' ∥ BB' ∥ CC', and P, Q, R be the intersections of A'D and BC, B'D and CA, C'D and AB, respectively. Then P, Q, R are collinear and the line passing through those three points is parallel to AA', BB', and CC'.*

Alternate Proof. Without loss of generality, we may assume that the circle under consideration is the unit circle. Let the points A, B, C, D be represented by the complex numbers α, β, γ, δ, respectively. Then

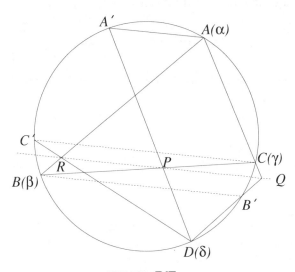

FIGURE 2.15

the equations of the parallel lines AA', BB', CC' are given by

$$z + k\overline{z} = \alpha + k\overline{\alpha}, \qquad z + k\overline{z} = \beta + k\overline{\beta}, \qquad z + k\overline{z} = \gamma + k\overline{\gamma},$$

respectively, where k is a suitable complex number with $|k| = 1$. It follows that the points A', B', C' are given by complex numbers $k\overline{\alpha}$, $k\overline{\beta}$, $k\overline{\gamma}$, respectively. Hence the intersection P of the lines BC and $A'D$ must be the solution of the system of simultaneous equations

$$z + \beta\gamma\overline{z} = \beta + \gamma, \qquad z + \delta k\overline{\alpha}\overline{z} = \delta + k\overline{\alpha}.$$

$$\therefore \ (\beta\gamma - k\delta\overline{\alpha})\overline{z} = \beta + \gamma - \delta - k\overline{\alpha}.$$

Multiplying both sides by α, we get

$$(\alpha\beta\gamma - k\delta)\overline{z} = \alpha\beta + \gamma\alpha - \delta\alpha - k.$$

Using the notations

$$\sigma_1 = \alpha + \beta + \gamma, \qquad \sigma_2 = \beta\gamma + \gamma\alpha + \alpha\beta, \qquad \sigma_3 = \alpha\beta\gamma,$$

the last equality can be rewritten as

$$(\sigma_3 - k\delta)\overline{z} = \sigma_2 - \frac{\sigma_3}{\alpha} - \delta\alpha - k.$$

Taking the complex conjugate and multiplying both sides by $k\delta\sigma_3$, we get

$$(k\delta - \sigma_3)z = \sigma_1 k\delta - k\delta\alpha - \frac{k\sigma_3}{\alpha} - \sigma_3\delta.$$

It follows from the last two equalities that

$$(k\delta - \sigma_3)(z + k\overline{z}) = k^2 + \sigma_1 k\delta - \sigma_2 k - \sigma_3\delta.$$

This is a relation that the point P must satisfy. But this relation is symmetric with respect to α, β, γ, and so the points Q and R also satisfy this relation. On the other hand, this is an equation of a line parallel to AA', BB', CC' (provided $k\delta - \sigma_3 \neq 0$). Hence, the points P, Q, R are collinear and the line passing through these three points is parallel to AA', BB', CC'.

In the case where $k\delta = \sigma_3$, we have

$$BC \parallel A'D, \qquad CA \parallel B'D, \qquad AB \parallel C'D,$$

and so the points P, Q, R all coincide with the point at infinity, and the conclusion is trivially true. $\square$

In concluding this section, we present another infinite sequence of theorems. But first we start from the case of four points.

THEOREM 2.6.4. *Let $A_1 A_2 A_3 A_4$ be a quadrangle inscribed in a circle, and P an arbitrary point on the circumcircle. Then the feet D_1, D_2, D_3, D_4 of the perpendiculars from P to the Simson lines of P with respect to*

$$\triangle A_2 A_3 A_4, \qquad \triangle A_1 A_3 A_4, \qquad \triangle A_1 A_2 A_4, \qquad \triangle A_1 A_2 A_3$$

*are collinear. This line is called the **Simson line** of the point P with respect to the quadrangle $A_1 A_2 A_3 A_4$.*

Proof. Without loss of generality, we may assume that the circumcircle of the quadrangle $A_1 A_2 A_3 A_4$ is the unit circle, and the complex numbers u_1, u_2, u_3, u_4, and u represent the points A_1, A_2, A_3, A_4, and P, respectively. Then the equation of the Simson line of the point $P(u)$ with respect to $\triangle A_2 A_3 A_4$ is

$$uz - u_2 u_3 u_4 \overline{z}$$
$$= \frac{1}{2} \left\{ u^2 + (u_2 + u_3 + u_4)u - (u_3 u_4 + u_4 u_2 + u_2 u_3) - \frac{u_2 u_3 u_4}{u} \right\}.$$

Hence the equation of the perpendicular from $P(u)$ to this Simson line is

$$uz + u_2 u_3 u_4 \overline{z} = u^2 + \frac{u_2 u_3 u_4}{u}.$$

Therefore, the intersection D_1 of these two lines is given by

$$2uz = \frac{1}{2} \left\{ 3u^2 + (u_2 + u_3 + u_4)u - (u_3 u_4 + u_4 u_2 + u_2 u_3) + \frac{u_2 u_3 u_4}{u} \right\}.$$

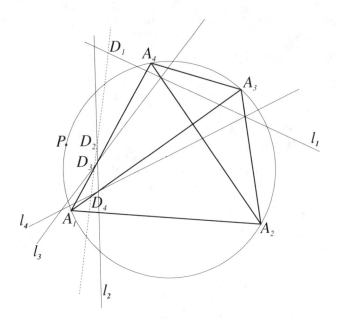

FIGURE 2.16

Using the notations

$$\sigma_1 = u_1 + u_2 + u_3 + u_4,$$

$$\sigma_2 = u_1u_2 + u_1u_3 + u_1u_4 + u_2u_3 + u_2u_4 + u_3u_4,$$

$$\sigma_3 = u_2u_3u_4 + u_1u_3u_4 + u_1u_2u_4 + u_1u_2u_3,$$

$$\sigma_4 = u_1u_2u_3u_4,$$

the last relation can be rewritten as

$$u^2 z = \frac{1}{4} \left\{ 3u^3 + (\sigma_1 - u_1)u^2 - (\sigma_2 - \sigma_1 u_1 + u_1^2)u + \frac{\sigma_4}{u_1} \right\}$$

$$= \frac{1}{4} \left\{ (3u^3 + \sigma_1 u^2 - \sigma_2 u) - (u_1 u^2 - \sigma_1 u_1 u + u_1^2 u) + \frac{\sigma_4}{u_1} \right\}.$$

Taking the complex conjugate of both sides, and using the relations

$$\overline{\sigma_1} = \frac{\sigma_3}{\sigma_4}, \qquad \overline{\sigma_2} = \frac{\sigma_2}{\sigma_4}, \qquad \overline{\sigma_4} = \frac{1}{\sigma_4},$$

we get

$$\overline{z} = \frac{u^2}{4} \left\{ \left(\frac{3}{u^3} + \frac{\sigma_3}{\sigma_4 u^2} - \frac{\sigma_2}{\sigma_4 u} \right) - \left(\frac{1}{u_1 u^2} - \frac{\sigma_3}{\sigma_4 u_1 u} + \frac{1}{u_1^2 u} \right) + \frac{u_1}{\sigma_4} \right\}.$$

$$\therefore \sigma_4 \overline{z} = \frac{1}{4} \left\{ \left(\frac{3\sigma_4}{u} + \sigma_3 - \sigma_2 u \right) - \left(\frac{\sigma_4}{u_1} - \frac{\sigma_3 u}{u_1} + \frac{\sigma_4 u}{u_1^2} \right) + u_1 u^2 \right\}.$$

Therefore,

$$u^2 z + \sigma_4 \overline{z} = \frac{1}{4} \left\{ \left(3u^3 + \sigma_1 u^2 - \sigma_2 u + \sigma_3 + \frac{3\sigma_4}{u} \right) \right.$$

$$+ u \left(\sigma_1 u_1 - u_1^2 - \sigma_2 + \frac{\sigma_3}{u_1} - \frac{\sigma_4}{u_1^2} \right) \right\}$$

$$= \frac{1}{4} \left\{ \left(3u^3 + \sigma_1 u^2 - \sigma_2 u + \sigma_3 + \frac{3\sigma_4}{u} \right) \right.$$

$$\left. - \frac{u}{u_1^2} \left(u_1^4 - \sigma_1 u_1^3 + \sigma_2 u_1^2 - \sigma_3 u_1 + \sigma_4 \right) \right\}.$$

Since u_1 is a root of

$$u^4 - \sigma_1 u^3 + \sigma_2 u^2 - \sigma_3 u + \sigma_4 = 0,$$

we obtain

$$u^2 z + \sigma_4 \overline{z} = \frac{1}{4} \left(3u^3 + \sigma_1 u^2 - \sigma_2 u + \sigma_3 + \frac{3\sigma_4}{u} \right).$$

This is a relation satisfied by D_1. However, by symmetry, it must also be satisfied by D_2, D_3, D_4. On the other hand, this is an equation of a straight line. Hence, the points D_1, D_2, D_3, D_4 are collinear. $\qquad \square$

Now, suppose there is a pentagon inscribed in a circle, and a point P on the circumcircle, . . .

2.7 The Cantor Theorems

Again we start from the following simple case.

THEOREM 2.7.1 (M. B. Cantor). *The three perpendiculars from the midpoints of the sides of a triangle to the tangents to the circumcircle at the opposite vertices meet at the center of the nine-point circle of the triangle.*

Proof. Without loss of generality, we may assume that $\triangle A_1 A_2 A_3$ is inscribed in the unit circle. Recall that the equation of the line passing through the points α and β on the unit circle is given by

$$z + \alpha\beta\bar{z} = \alpha + \beta.$$

Since the tangent to the unit circle at the point α is the particular case when α and β coincide, the equation of the tangent at α is

$$z + \alpha^2\bar{z} = 2\alpha.$$

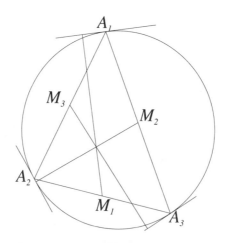

FIGURE 2.17

Let A_1, A_2, A_3 be represented by the complex numbers u_1, u_2, u_3, respectively. Then the equation of the tangent at A_1 is

$$z + u_1^2 \bar{z} = 2u_1.$$

Hence the equation of the perpendicular from the midpoint $M_1 \left(\dfrac{u_2 + u_3}{2} \right)$ of the side $A_2 A_3$ to this tangent line is

$$z - u_1^2 \bar{z} = \frac{1}{2} \left\{ (u_2 + u_3) - u_1^2 (\bar{u}_2 + \bar{u}_3) \right\}.$$

Substituting the center $\frac{1}{2}(u_1 + u_2 + u_3)$ of the nine-point circle of $\triangle A_1 A_2 A_3$ into the left-hand side of this equation, we get

$$\frac{1}{2} \left\{ (u_1 + u_2 + u_3) - u_1^2 (\bar{u}_1 + \bar{u}_2 + \bar{u}_3) \right\}$$

$$= \frac{1}{2} \left\{ (u_2 + u_3) - u_1^2 (\bar{u}_2 + \bar{u}_3) \right\} \quad (\because u_1 \bar{u}_1 = 1),$$

which coincides with the right-hand side of the equation. Therefore, the center of the nine-point circle satisfies the equation of the perpendicular from M_1 to the tangent at A_1. Similarly, the center of the nine-point circle is on the perpendiculars from M_2 and M_3 to the tangents at the respective opposite vertices. $\square$

THEOREM 2.7.2 (M. B. Cantor). *Let n points be given on a circle. From the centroid of $n - 1$ of these points, drop a perpendicular to the tangent to the circle at the remaining point. Then these n perpendiculars meet at a point.*

Proof. The proof is practically the same as the previous one. Let $u_1, u_2, \ldots, u_n$ be n points on the unit circle. Then the equation of the tangent at u_1 is $z + u_1^2 \bar{z} = 2u_1$, hence the equation of the perpendicular from the centroid

$$\frac{1}{n-1}(u_2 + u_3 + \cdots + u_n) = \frac{(\sigma_1 - u_1)}{n - 1} \quad (\sigma_1 = u_1 + u_2 + \cdots + u_n),$$

of the points $u_2, u_3, \ldots, u_n$ to this tangent is

$$z - u_1^2 \bar{z} = \frac{1}{n-1} \left\{ (u_2 + u_3 + \cdots + u_n) - u_1^2 (\bar{u}_2 + \bar{u}_3 + \cdots + \bar{u}_n) \right\}$$

$$= \frac{1}{n-1} \left\{ (\sigma_1 - u_1) - u_1^2 (\bar{\sigma}_1 - \bar{u}_1) \right\}$$

$$= \frac{1}{n-1} (\sigma_1 - u_1^2 \bar{\sigma}_1).$$

It is obvious that the point

$$\frac{1}{n-1} (u_1 + u_2 + \cdots + u_n) = \frac{\sigma_1}{n-1}$$

satisfies this equation. $\qquad\qquad\qquad\qquad\qquad\qquad\qquad\qquad\square$

We now embark on yet another infinite sequence of theorems discovered by M. B. Cantor (1829–1920).

THEOREM 2.7.3. *Let A_1, A_2, A_3, A_4, P_1, P_2 be six cocyclic points. Then the four intersections of the four pairs of Simson lines of the points P_1 and P_2 with respect to*

$$\triangle A_2 A_3 A_4, \qquad \triangle A_1 A_3 A_4, \qquad \triangle A_1 A_2 A_4, \qquad \triangle A_1 A_2 A_3,$$

*are collinear. This line is called the **Cantor line** of the pair of points P_1 and P_2 with respect to the quadrangle $A_1 A_2 A_3 A_4$.*

Proof. Without loss of generality, we may assume that all of these six points are on the unit circle, and that they are represented by complex numbers u_1, u_2, u_3, u_4 and t_1, t_2, respectively. Then the equations of the Simson lines of the points $P_1(t_1)$ and of $P_2(t_2)$ with respect to $\triangle A_2 A_3 A_4$ are given by

$$t_1 z - u_2 u_3 u_4 \bar{z}$$

$$= \frac{1}{2} \left\{ t_1^2 + (u_2 + u_3 + u_4) t_1 - (u_3 u_4 + u_4 u_2 + u_2 u_3) - \frac{u_2 u_3 u_4}{t_1} \right\},$$

$$t_2 z - u_2 u_3 u_4 \bar{z}$$

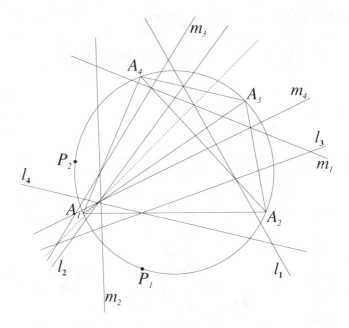

FIGURE 2.18

$$= \frac{1}{2} \left\{ t_2^2 + (u_2 + u_3 + u_4)t_2 - (u_3 u_4 + u_4 u_2 + u_2 u_3) - \frac{u_2 u_3 u_4}{t_2} \right\}.$$

Hence their intersection is given by

$$z = \frac{1}{2} \left\{ t_1 + t_2 + (u_2 + u_3 + u_4) + \frac{u_2 u_3 u_4}{t_1 t_2} \right\}.$$

Setting

$$\sigma_1 = u_1 + u_2 + u_3 + u_4,$$

$$\sigma_2 = u_1 u_2 + u_1 u_3 + u_1 u_4 + u_2 u_3 + u_2 u_4 + u_3 u_4,$$

$$\sigma_3 = u_2 u_3 u_4 + u_1 u_3 u_4 + u_1 u_2 u_4 + u_1 u_2 u_3,$$

$$\sigma_4 = u_1 u_2 u_3 u_4,$$

the above relation can be rewritten as

$$z = \frac{1}{2}\left\{ t_1 + t_2 + \sigma_1 - u_1 + \frac{\sigma_4}{t_1 t_2 u_1} \right\},$$

and so

$$\bar{z} = \frac{1}{2}\left\{ \frac{1}{t_1} + \frac{1}{t_2} + \frac{\sigma_3}{\sigma_4} - \frac{1}{u_1} + \frac{t_1 t_2 u_1}{\sigma_4} \right\}.$$

Eliminating u_1 from the last two relations, we get

$$t_1 t_2 z + \sigma_4 \bar{z} = \frac{1}{2}\left\{ (t_1 + t_2)t_1 t_2 + \sigma_1 t_1 t_2 + \sigma_3 + \frac{\sigma_4(t_1 + t_2)}{t_1 t_2} \right\}.$$

This is a relation that must be satisfied by the intersection of the Simson line of P_1 and that of P_2 with respect to $\triangle A_2 A_3 A_4$. However, this relation is symmetric with respect to u_1, u_2, u_3, u_4, and hence is satisfied by the intersections of the pairs of the Simson lines of P_1 and P_2 with respect to $\triangle A_1 A_3 A_4$, $\triangle A_1 A_2 A_4$, $\triangle A_1 A_2 A_3$. On the other hand, this relation is the equation of a line. Therefore, these four intersections are collinear.   □

THEOREM 2.7.4. *Let A_1, A_2, A_3, A_4, P_1, P_2, P_3 be seven cocyclic points. Then the three Cantor lines of the three pairs of points P_2 and P_3, P_3 and P_1, P_1 and P_2 with respect to the quadrangle $A_1 A_2 A_3 A_4$ meet at a point. This point is called the **Cantor point** of the triple of points P_1, P_2, P_3 with respect to the quadrangle $A_1 A_2 A_3 A_4$.*

Proof. As before, let the circle under consideration be the unit circle, and u_1, u_2, u_3, u_4, t_1, t_2, t_3 the complex numbers representing the points A_1, A_2, A_3, A_4, P_1, P_2, P_3, respectively. Then the equations of the Cantor lines of the pairs of points P_2 and P_3, P_3 and P_1 with respect to the quadrangle $A_1 A_2 A_3 A_4$ are given by

$$t_2 t_3 z + \sigma_4 \bar{z} = \frac{1}{2}\left\{ (t_2 + t_3)t_2 t_3 + \sigma_1 t_2 t_3 + \sigma_3 + \frac{\sigma_4(t_2 + t_3)}{t_2 t_3} \right\},$$

$$t_3 t_1 z + \sigma_4 \bar{z} = \frac{1}{2}\left\{ (t_3 + t_1)t_3 t_1 + \sigma_1 t_3 t_1 + \sigma_3 + \frac{\sigma_4(t_3 + t_1)}{t_3 t_1} \right\}.$$

Hence their intersection is given by

$$z = \frac{1}{2} \left\{ t_1 + t_2 + t_3 + \sigma_1 - \frac{\sigma_4}{t_1 t_2 t_3} \right\}.$$

However, this expression is symmetric with respect to t_1, t_2, t_3, and hence the Cantor line of the pair of points P_1 and P_2 with respect to the quadrangle $A_1 A_2 A_3 A_4$ also passes through this point. Thus this point must be the Cantor point of the triple of points P_1, P_2, P_3 with respect to the quadrangle $A_1 A_2 A_3 A_4$. $\qquad\square$

THEOREM 2.7.5. *Let A_1, A_2, A_3, A_4, A_5, P_1, P_2, P_3, be eight cocyclic points. Then the five Cantor points of the triple of points P_1, P_2, P_3 with respect to the quadrangles*

$$A_2 A_3 A_4 A_5, \ A_1 A_3 A_4 A_5, \ A_1 A_2 A_4 A_5, \ A_1 A_2 A_3 A_5, \ A_1 A_2 A_3 A_4$$

*are collinear. This line is called the **Cantor line** of the triple of points P_1, P_2, P_3 with respect to the pentagon $A_1 A_2 A_3 A_4 A_5$.*

Proof. As before, let the circle under consideration be the unit circle, and u_1, u_2, u_3, u_4, u_5, t_1, t_2, t_3 the complex numbers representing the points A_1, A_2, A_3, A_4, A_5, P_1, P_2, P_3, respectively. Then the Cantor point of the triple of points P_1, P_2, P_3 with respect to the the quadrangle $A_2 A_3 A_4 A_5$ is given by

$$z = \frac{1}{2} \left\{ t_1 + t_2 + t_3 + u_2 + u_3 + u_4 + u_5 - \frac{u_2 u_3 u_4 u_5}{t_1 t_2 t_3} \right\}.$$

Setting

$$\sigma_1 = u_1 + u_2 + u_3 + u_4 + u_5,$$

$$\sigma_2 = u_1 u_2 + u_1 u_3 + u_1 u_4 + u_1 u_5 + u_2 u_3$$
$$+ u_2 u_4 + u_2 u_5 + u_3 u_4 + u_3 u_5 + u_4 u_5,$$

$$\sigma_3 = u_1 u_2 u_3 + u_1 u_2 u_4 + u_1 u_2 u_5 + u_1 u_3 u_4 + u_1 u_3 u_5$$
$$+ u_1 u_4 u_5 + u_2 u_3 u_4 + u_2 u_3 u_5 + u_2 u_4 u_5 + u_3 u_4 u_5,$$

$$\sigma_4 = u_2 u_3 u_4 u_5 + u_1 u_3 u_4 u_5 + u_1 u_2 u_4 u_5 + u_1 u_2 u_3 u_5 + u_1 u_2 u_3 u_4,$$

$$\sigma_5 = u_1 u_2 u_3 u_4 u_5,$$

the above relation can be rewritten as

$$z = \frac{1}{2} \left\{ t_1 + t_2 + t_3 + \sigma_1 - u_1 - \frac{\sigma_5}{u_1 t_1 t_2 t_3} \right\}.$$

$$\therefore \ \bar{z} = \frac{1}{2} \left\{ \frac{1}{t_1} + \frac{1}{t_2} + \frac{1}{t_3} + \frac{\sigma_4}{\sigma_5} - \frac{1}{u_1} - \frac{u_1 t_1 t_2 t_3}{\sigma_5} \right\}.$$

Eliminating u_1 from these two relations, we get

$$t_1 t_2 t_3 z - \sigma_5 \bar{z}$$

$$= \frac{1}{2} \left\{ (t_1 + t_2 + t_3) t_1 t_2 t_3 + \sigma_1 t_1 t_2 t_3 - \sigma_4 - \frac{\sigma_5 (t_2 t_3 + t_3 t_1 + t_1 t_2)}{t_1 t_2 t_3} \right\}.$$

This is a relation that must be satisfied by the Cantor point of the triple of points P_1, P_2, P_3 with respect to the quadrangle $A_2 A_3 A_4 A_5$. However, this relation is symmetric with respect to A_1, A_2, A_3, A_4, A_5, and hence must also be satisfied by the Cantor points of the triple P_1, P_2, P_3 with respect to the quadrangles $A_1 A_3 A_4 A_5$, $A_1 A_2 A_4 A_5$, $A_1 A_2 A_3 A_5$, $A_1 A_2 A_3 A_4$. On the other hand, this is an equation of a line. Hence these five Cantor points must be collinear.   □

Now, suppose we have nine cocyclic points A_1, A_2, A_3, A_4, A_5, P_1, P_2, P_3, P_4, then ...

2.8 The Feuerbach Theorem

Let H be the orthocenter of $\triangle ABC$. Noting that the nine-point circle of $\triangle ABC$ is also the nine-point circle of $\triangle HBC$, $\triangle HCA$, and $\triangle HAB$, the following theorem of 1822 attributed to K. W. Feuerbach (1800–1834), a high-school teacher in Erlangen, Germany, is truly remarkable.

THEOREM 2.8.1 (Feuerbach). *The nine-point circle of a triangle is tangent to the incircle and the three excircles.* (See Figure 2.19.)

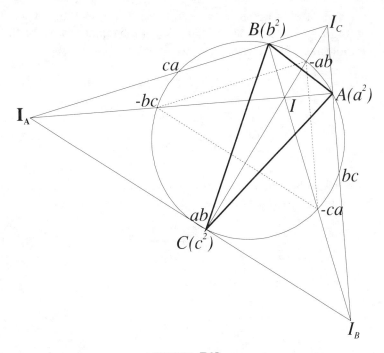

FIGURE 2.19

Proof. Without loss of generality, we may assume that $\triangle ABC$ is inscribed in the unit circle. To avoid the cumbersome square root signs, let us assume that the vertices A, B, C are represented by complex numbers a^2, b^2, c^2, respectively.

The (interior) angle bisector at the vertex A passes through the midpoint of $\overset{\frown}{BC}$, which does not contain the vertex A, while the exterior angle bisector at the vertex A passes through the midpoint of $\overset{\frown}{BC}$, which contains the vertex A. Let the latter be bc, then the former must be $-bc$. Similarly, let the midpoint of $\overset{\frown}{CA}$, which contains the vertex B be ca and that of $\overset{\frown}{CA}$, which does not contain the vertex B be $-ca$; and the midpoint of $\overset{\frown}{AB}$, which contains the vertex C be ab and that of $\overset{\frown}{CA}$, which does not contain the vertex C be $-ab$. (Note that this can always be achieved by changing the sign(s) of a, b, or c if necessary. For example, suppose A, B, C are situated counterclockwise on the unit circle with A at the point $z = 1$. Choose a, b, c such that $a = -1$, $0 < \arg b < \pi$, $-\pi < \arg c < 0$.) Then the equations of the three (interior) angle bisectors are

$$z - a^2 b c \bar{z} = a^2 - bc,$$

$$z - ab^2 c \bar{z} = b^2 - ca,$$

$$z - abc^2 \bar{z} = c^2 - ab,$$

respectively. Solving a system of simultaneous equations obtained by choosing any two of these three equations, we get

$$z = -(bc + ca + ab).$$

As usual, if we denote

$$\sigma_1 = a + b + c, \qquad \sigma_2 = bc + ca + ab, \qquad \sigma_3 = abc,$$

then the intersection is $z = -\sigma_2$. Clearly, this also satisfies the remaining equation (by symmetry). We have shown that the three angle bisectors of a triangle meet at a point; this point is called the *incenter* of the triangle.

But I: $-\sigma_2 = -bc - ca - ab$ is also the orthocenter of the triangle with vertices at $-bc$, $-ca$, $-ab$. In hindsight, this is obvious if we compare the equation of the (interior) angle bisector at the vertex A with that of the line joining the points $-ab$ and $-ca$:

$$z + a^2 b c \bar{z} = -ca - ab.$$

This is also true for the other two angle bisectors.

Similar observation tells us that the excenters I_A, I_B, I_C are the orthocenters of the triangles with vertices at

$$
\begin{array}{ccc}
-bc, & ab, & ca; \\
-ca, & bc, & ab; \\
-ab, & ca, & bc;
\end{array}
$$

respectively. Since all these triangles are inscribed in the unit circle, we have

$$I_A \; : \; -bc + ab + ca,$$
$$I_B \; : \; -ca + bc + ab,$$
$$I_C \; : \; -ab + ca + bc.$$

We now compute the distance d between the incenter I and the center of the nine-point circle of $\triangle ABC$.

$$d = \left| \frac{1}{2}(a^2 + b^2 + c^2) + \sigma_2 \right|$$

$$= \frac{1}{2} \left| a^2 + b^2 + c^2 + 2(bc + ca + ab) \right|$$

$$= \frac{1}{2} \left| (a + b + c)^2 \right| = \frac{1}{2} \sigma_1 \overline{\sigma}_1 = \frac{\sigma_1 \sigma_2}{2\sigma_3}.$$

We know the radius of the nine-point circle is $\frac{1}{2}$. Let us compute the radius r of the incircle. The equation of the line BC is

$$z + b^2 c^2 \overline{z} = b^2 + c^2,$$

and so the equation of the perpendicular from the incenter $I(-\sigma_2)$ to BC is

$$z - b^2 c^2 \overline{z} = -\sigma_2 + b^2 c^2 \overline{\sigma}_2 = -\sigma_2 + \frac{bc\sigma_1}{a}.$$

Hence the foot of this perpendicular is given by

$$z = \frac{1}{2} \left(b^2 + c^2 - \sigma_2 + \frac{bc\sigma_1}{a} \right).$$

It follows that the radius r of the incircle is

$$r = \left| \frac{1}{2} \left(b^2 + c^2 - \sigma_2 + \frac{bc\sigma_1}{a} \right) + \sigma_2 \right|$$

$$= \frac{1}{2} \left| a(b^2 + c^2) + a\sigma_2 + bc\sigma_1 \right|$$

$$= \frac{1}{2} \left| \sigma_1 \sigma_2 - \sigma_3 \right|$$

$$= \left| \frac{\sigma_1 \sigma_2}{2\sigma_3} - \frac{1}{2} \right| = \left| d - \frac{1}{2} \right|,$$

where we used the fact $|a| = 1 = |\sigma_3|$. It is simple to verify that $d \leq \frac{1}{2}$, therefore, we get

$$r = \frac{1}{2} - d; \quad \text{i.e.,} \quad d = \frac{1}{2} - r,$$

which shows that the nine-point circle and the incircle are tangent to each other internally.

To prove that the nine-point circle is tangent to the excircle I_A, say, we have merely to replace a by $-a$ and repeat the above argument. $\square$

Alternate Proof. In the proof which appeared in P. J. Davis's *The Schwarz Function and its Applications* [Mathematical Association of America, Washington, D.C., 1974, pp.16–18], the unit circle is taken not as the circumcircle, but as the incircle of $\triangle ABC$. Let the points of tangency of the three sides BC, CA, AB with the unit circle be α, β, γ, respectively.

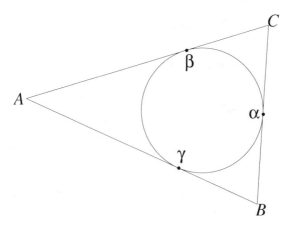

FIGURE 2.20

Then the equations of these three sides are

$$z + \alpha^2 \bar{z} = 2\alpha, \qquad z + \beta^2 \bar{z} = 2\beta, \qquad z + \gamma^2 \bar{z} = 2\gamma.$$

Solving the systems of simultaneous equations consisting of two of these three equations, we obtain that the vertices A, B, C are given by the complex numbers

$$A: \frac{2\beta\gamma}{\beta + \gamma}, \qquad B: \frac{2\gamma\alpha}{\gamma + \alpha}, \qquad C: \frac{2\alpha\beta}{\alpha + \beta}.$$

Therefore, the midpoint L of the side BC is

$$\alpha \left(\frac{\gamma}{\gamma + \alpha} + \frac{\beta}{\alpha + \beta} \right) = \frac{(2\beta\gamma + \gamma\alpha + \alpha\beta)(\alpha\beta + \gamma\alpha)}{(\gamma + \alpha)(\alpha + \beta)(\beta + \gamma)}$$

$$= \frac{(\sigma_2 + \beta\gamma)(\sigma_2 - \beta\gamma)}{\sigma_1\sigma_2 - \sigma_3} = \frac{\sigma_2^2 - (\beta\gamma)^2}{\sigma_1\sigma_2 - \sigma_3}$$

$$= \frac{\sigma_2^2}{\sigma_1\sigma_2 - \sigma_3} - \frac{\sigma_3^2}{\alpha^2(\sigma_1\sigma_2 - \sigma_3)},$$

where

$$\sigma_1 = \alpha + \beta + \gamma, \qquad \sigma_2 = \beta\gamma + \gamma\alpha + \alpha\beta, \qquad \sigma_3 = \alpha\beta\gamma.$$

Similarly, the midpoints M, N of the sides CA, AB are

$$M: \frac{\sigma_2^2}{\sigma_1\sigma_2 - \sigma_3} - \frac{\sigma_3^2}{\beta^2(\sigma_1\sigma_2 - \sigma_3)}, \qquad N: \frac{\sigma_2^2}{\sigma_1\sigma_2 - \sigma_3} - \frac{\sigma_3^2}{\gamma^2(\sigma_1\sigma_2 - \sigma_3)}.$$

Let K be the point represented by $\dfrac{\sigma_2^2}{\sigma_1\sigma_2 - \sigma_3}$, then clearly

$$\overline{KL} = \overline{KM} = \overline{KN} = \frac{1}{|\sigma_1\sigma_2 - \sigma_3|},$$

so K must be the center of the nine-point circle, and the radius is $1/|\sigma_1\sigma_2 - \sigma_3|$. Hence the equation of the nine-point circle is

$$\left| z - \frac{\sigma_2^2}{\sigma_1\sigma_2 - \sigma_3} \right| = \frac{1}{|\sigma_1\sigma_2 - \sigma_3|};$$

i.e.,

$$\left(z - \frac{\sigma_2^2}{\sigma_1\sigma_2 - \sigma_3} \right)\left(\overline{z} - \frac{\sigma_1^2}{\sigma_1\sigma_2 - \sigma_3} \right) = \left(\frac{\sigma_3^2}{\sigma_1\sigma_2 - \sigma_3} \right)^2,$$

where we have used the relations

$$\overline{\sigma}_1 = \frac{\sigma_2}{\sigma_3}, \qquad \overline{\sigma}_2 = \frac{\sigma_1}{\sigma_3}, \qquad \overline{\sigma}_3 = \frac{1}{\sigma_3}.$$

Rewriting the last equation, we get

$$z\overline{z} - \left(\frac{\sigma_1^2 z}{\sigma_1\sigma_2 - \sigma_3} + \frac{\sigma_2^2 \overline{z}}{\sigma_1\sigma_2 - \sigma_3} \right) + \frac{\sigma_1\sigma_2 + \sigma_3}{\sigma_1\sigma_2 - \sigma_3} = 0.$$

Solving the system of simultaneous equations consisting of the last equation (the equation of the nine-point circle) and the equation of the incircle $z\overline{z} = 1$, we obtain

$$\sigma_1^2 z^2 - 2\sigma_1\sigma_2 z + \sigma_2^2 = 0;$$

i.e.,

$$(\sigma_1 z - \sigma_2)^2 = 0.$$

It follows that if $\sigma_1 \neq 0$, then the system of equations of the two circles has a double root, which means that these two circles are tangent to each other. If $\sigma_1 = 0$, then the triangle is equilateral (see exercise 1.35), and the incircle and the nine-point circle coincide.

The above computation is valid without change even if two of α, β, γ are on the extensions of the respective sides (i.e., considering the unit circle as an excircle), hence our proof is complete.　　　　□

2.9 The Morley Theorem

The following theorem discovered by Frank Morley (1860–1934) around the turn of this century certainly qualifies as one of the most beautiful theorems in mathematics.

THEOREM 2.9.1 (Morley). *The intersections of the adjacent pairs of angle trisectors of an arbitrary triangle are the vertices of an equilateral triangle.*

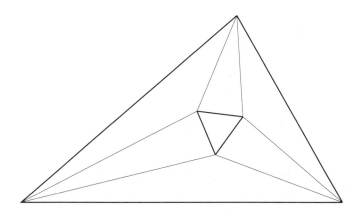

FIGURE 2.21

Before we start a proof of this theorem, we need the following

LEMMA 2.9.2. *Suppose t_1, t_2, t_3, t_4 are points on the unit circle. Then (the extensions of) the chords joining the points t_1, t_2, and t_3, t_4 meet at*

$$z = \frac{\bar{t}_1 + \bar{t}_2 - \bar{t}_3 - \bar{t}_4}{\bar{t}_1\bar{t}_2 - \bar{t}_3\bar{t}_4}.$$

Proof. We know the equations of the lines passing through the points t_1 and t_2, and that of t_3 and t_4 are

$$z + t_1t_2\bar{z} = t_1 + t_2, \qquad z + t_3t_4\bar{z} = t_3 + t_4,$$

respectively. Hence the intersection of these two lines is

$$z = \frac{\bar{t}_1 + \bar{t}_2 - \bar{t}_3 - \bar{t}_4}{\bar{t}_1\bar{t}_2 - \bar{t}_3\bar{t}_4}.$$

□

Proof of the theorem.   Without loss of generality, we may assume that $\triangle ABC$ is inscribed in the unit circle, and that vertex A is at point 1. Let

$$\angle AOB = 3\gamma \qquad \left(0 < \gamma < \frac{2\pi}{3}\right),$$

$$\angle AOC = 3\beta \qquad \left(-\frac{2\pi}{3} < \beta < 0\right),$$

$$\angle BOC = 3\alpha \qquad \left(\alpha = \frac{2\pi}{3} + \beta - \gamma > 0\right).$$

Then the arguments of the points trisecting $\overset{\frown}{BC}$ (not containing the point A) are

$$\alpha + 3\gamma = \beta + 2\gamma + \frac{2\pi}{3},$$

and

$$2\alpha + 3\gamma = 2\beta + \gamma + \frac{4\pi}{3}.$$

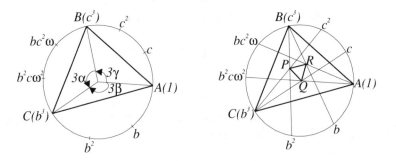

FIGURE 2.22

Therefore, if we set the points trisecting $\overset{\frown}{AB}$ and $\overset{\frown}{AC}$ to be c, c^2, and b, b^2, respectively, then B, C are c^3 and b^3, and the points trisecting $\overset{\frown}{BC}$ are given by $bc^2\omega$ and $b^2c\omega^2$, where $\omega^2 + \omega + 1 = 0$.

Let $P(\lambda)$, $Q(\mu)$, $R(\nu)$ be the intersections of the adjacent trisectors of the angles at B and C, C and A, A and B, respectively. Then, by the lemma,

$$\lambda = \frac{b^{-2} + c^{-3} - b^{-3} - c^{-2}}{b^{-2}c^{-3} - b^{-3}c^{-2}} = \frac{bc^3 + b^3 - c^3 - b^3c}{b - c}$$

$$= \frac{(b - c)(b^2 + bc + c^2) - bc(b^2 - c^2)}{b - c} = (b^2 + bc + c^2) - bc(b + c),$$

$$\mu = \frac{1 + b^{-2}c^{-1}\omega^{-2} - b^{-3} - c^{-1}}{b^{-2}c^{-1}\omega^{-2} - b^{-3}c^{-1}} = \frac{b^3c + b\omega - c - b^3}{b\omega - 1}$$

$$= \frac{c(b^3 - 1) - b(b^2 - \omega)}{\omega(b - \omega^2)} = \omega^2 \left\{ c(b^2 + b\omega^2 + \omega) - b(b + \omega^2) \right\},$$

$$\nu = \frac{1 + b^{-1}c^{-2}\omega^{-1} - b^{-1} - c^{-3}}{b^{-1}c^{-2}\omega^{-1} - b^{-1}c^{-3}} = \frac{bc^3 + c\omega^2 - c^3 - b}{c\omega^2 - 1}$$

$$= \frac{b(c^3 - 1) - c(c^2 - \omega^2)}{\omega^2(c - \omega)} = \omega \left\{ b(c^2 + c\omega + \omega^2) - c(c + \omega) \right\}.$$

$$\therefore \ \lambda + \omega\mu + \omega^2\nu = b^2 + bc + c^2 - b^2c - bc^2$$

$$+ b^2c + bc\omega^2 + c\omega - b^2 - b\omega^2$$

$$+ bc^2 + bc\omega + b\omega^2 - c^2 - c\omega$$

$$= 0,$$

and so we are done.

It is well known that if the (interior) angle trisectors are replaced by the exterior angle trisectors, then the conclusion of the Morley theorem is still valid. How can the proof be modified to cover such a case? For this purpose, first observe that the interior and exterior angle trisectors form an angle of $\frac{\pi}{3}$ between them. Since the central angle is twice

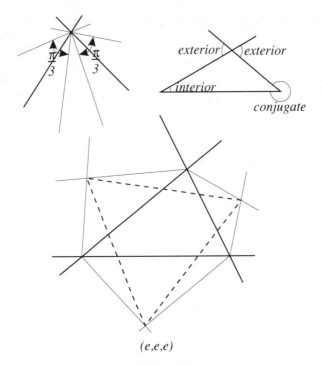

exterior   *exterior*

interior

conjugate

(e,e,e)

FIGURE 2.23

the inscribed angle, P (which was the intersection of the lines passing through b^3, c^2, and c^3, b^2) becomes the intersection of the lines through b^3, $c^2\omega$, and c^3, $b^2\omega^2$; Q (which was the intersection of the lines passing through b^3, c, and 1, $b^2c\omega^2$) becomes the intersection of the lines through b^3, $c\omega^2$, and 1, b^2c; while R (which was the intersection of the lines passing through c^3, b, and 1, $bc^2\omega$) becomes the intersection of the lines through c^3, $b\omega$, and 1, bc^2.

But this amounts merely to replacing b by $b\omega$, and c by $c\omega^2$ in the original proof, and the original computation is clearly valid under this transformation! So the case of trisecting the exterior angles is established with no extra cost.

But are there any other valid transformations? To pursue this question, let us introduce the transformations:

$$T_b(b) = b\omega, \quad T_b(c) = c; \qquad\qquad T_c(b) = b, \quad T_c(c) = c\omega;$$

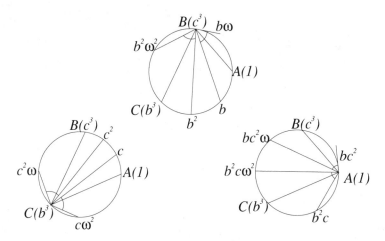

FIGURE 2.24

i.e., T_b changes b to $b\omega$ but keeps c unchanged, while T_c keeps b unchanged but changes c to $c\omega$. For example,

$$T_b T_c^2(b) = b\omega, \qquad T_b T_c^2(c) = c\omega^2.$$

Thus, the case of trisecting the exterior angles becomes simply applying the transformation $T_b T_c^2$ ($= T_c^2 T_b$). But it is obvious that the original proof is valid under other transformations, say $T_b^2 T_c$ ($= T_c T_b^2$). What is the geometric meaning of applying the transformation $T_b^2 T_c$; i.e., replacing b by $b\omega^2$, and c by $c\omega$? It is easy to see this is precisely the case of replacing all the interior angle trisectors by the conjugate angle trisectors. So the conclusion of the Morley theorem is valid in this case too.

Since $T_b^3 =$ the identity transformation $= T_c^3$, we have the following 9 cases, where (i, e, c), for example, indicates taking the interior, exterior, conjugate angle trisectors at the vertices A, B, C, respectively.

	I	T_b	T_b^2
I	(i,i,i)	(c,e,i)	(e,c,i)
T_c	(e,i,c)	(i,e,c)	(c,c,c)
T_c^2	(c,i,e)	(e,e,e)	(i,c,e)

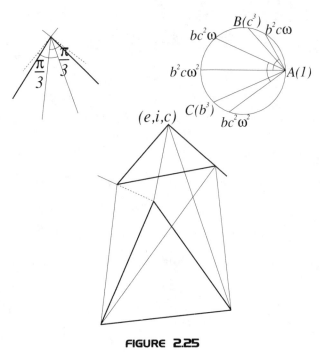

FIGURE 2.25

Actually, we can do more. In the above 9 cases, we have fixed *each* of the three vertices of the triangle invariant, but all we need is to keep the triangle itself invariant. Hence, we introduce the transformation S:

$$S(b) = c, \qquad S(c) = b.$$

In other words, S switches the roles of b and c. The transformation S may not have any effect on the trisectors at the vertices B and C (they simply switch roles only), but it does replace the interior angle trisectors with conjugate angle trisectors at the vertex A. Since $S^2 =$ the identity transformation, again we have 9 cases—listed in the following table.

S	I	T_b	T_b^2
I	(c, i, i)	(i, i, c)	(e, i, e)
T_c	(e, e, i)	(c, e, c)	(i, e, e)
T_c^2	(i, c, i)	(e, c, c)	(c, c, e)

For example,

$$ST_b(b) = c\omega, \qquad ST_b(c) = b;$$

$$ST_b^2 T_c(b) = c\omega^2, \qquad ST_b^2 T_c(c) = b\omega,$$

which result in the types (i, i, c) and (i, e, e), respectively.
Summing up, we have :

(1) one permutation for each of (i, i, i), (e, e, e), (c, c, c);

(2) 6 permutations of (i, e, c);

(3) 3 permutations for each of (i, e, e), (e, c, c), (c, i, i).

Hence the total of 18 equilateral triangles with 7 different types.

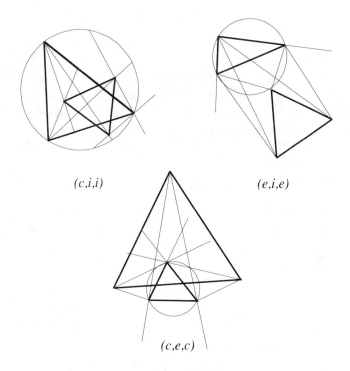

(c,i,i) (e,i,e)

(c,e,c)

FIGURE 2.26

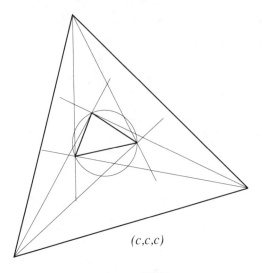

(c,c,c)

FIGURE 2.27

The missing types such as (i,c,c), (e,i,i), (c,e,e) do not result in equilateral triangles.

It is interesting to note that to construct these 18 equilateral triangles, we need to draw 18 angle trisectors, and each of these 18 trisectors passes through exactly 3 vertices (of the 18 equilateral triangles), and each of these 27 vertices is used exactly twice.

Exercises

1. (a) On each side of an arbitrary quadrangle, draw a square externally. Show that the two line segments joining the centers of the opposite squares are perpendicular to each other and of the same length.

   (b) Discuss the case in which the squares are drawn internally.

   (c) What if one of the sides of a given quadrangle degenerates to a point?

2. (a) On each side of an arbitrary parallelogram draw a square externally. Show that their centers form the vertices of a square.

(b) What if the squares are drawn internally?

3. (a) Suppose there are constants $a, b \in \mathbb{C}$, $a \neq 0$, satisfying

$$w_k = az_k + b \qquad (k = 1, 2, 3).$$

Show that $\triangle z_1 z_2 z_3 \sim \triangle w_1 w_2 w_3$.

(b) Is the converse true?

(c) Interpret the condition geometrically.

4. (a) Show that $\triangle z_1 z_2 z_3 \sim \triangle w_1 w_2 w_3$ if and only if there are constants $\alpha, \beta, \gamma \in \mathbb{C}$, not all zero, such that

$$\alpha z_1 + \beta z_2 + \gamma z_3 = 0, \qquad \alpha w_1 + \beta w_2 + \gamma w_3 = 0, \qquad \alpha + \beta + \gamma = 0.$$

(b) What if $\triangle z_1 z_2 z_3 \sim \triangle w_1 w_2 w_3$ (reversed)?

(c) Suppose four points z_1, z_2, z_3, z_4 satisfy

$$z_1 + i z_2 - z_3 - i z_4 = 0.$$

What can be said about the quadrangle $z_1 z_2 z_3 z_4$?

5. (a) (W. H. Echols) Suppose $\triangle ABC$ and $\triangle A'B'C'$ are equilateral triangles with the same orientation. Show that the midpoints of the segments AA', BB', CC' are the vertices of an equilateral triangle.

(b) What if the 'midpoints' in (a) are replaced by points that divide AA', BB', CC' into a fixed ratio, say $m : n$?

(c) What if $\triangle ABC \sim \triangle A'B'C'$ instead of 'equilateral'?

6. (a) (W. H. Echols) Suppose $\triangle ABC$, $\triangle DEF$, $\triangle GHI$ are equilateral with the same orientation. Let P, Q, R be the centroids of $\triangle ADG$, $\triangle BEH$, $\triangle CFI$, respectively. Show that $\triangle PQR$ is equilateral.

(b) What if 'equilateral' in (a) is replaced by 'similar'?

(c) What if 'centroids' are replaced by 'the points with the same barycentric coordinates in the respective triangle'?

7. (J. Petersen–P. H. Schoute) Suppose $\triangle ABC \sim \triangle A_1B_1C_1$ and $\triangle AA_1A_2 \sim \triangle BB_1B_2 \sim \triangle CC_1C_2$. Show that

$$\triangle A_2B_2C_2 \sim \triangle ABC.$$

8. Let A', B', C' be the points that divide the sides BC, CA, AB of a triangle ABC into a fixed ratio, say $m : n$. Show that

$$\triangle A'B'C' \sim \triangle ABC$$

if and only if either $\triangle ABC$ is equilateral (in this case $m : n$ is arbitrary), or $m : n = 1 : 1$ (in this case the shape of $\triangle ABC$ is arbitrary).

9. Given a triangle ABC, draw external squares $ABDE$ and $ACFG$ on the sides AB and AC, respectively.

(a) Let M be the midpoint of the third side BC. Show that

$$EG \perp AM \quad \text{and} \quad \overline{EG} = 2\overline{AM}.$$

(b) Let H be the foot of the perpendicular from the vertex A to the side BC. Show that the extension of AH bisects EG.

10. On the sides AB, BC of an arbitrary parallelogram $ABCD$, external equilateral triangles AEB and BFC are constructed. Show that $\triangle DEF$ is equilateral.

11. On each of the two opposite sides of an arbitrary quadrangle draw an external equilateral triangle, and on each of the remaining opposite sides draw an internal equilateral triangle. Show that the four new vertices are the vertices of a parallelogram.

12. (a) Show that the Napoleon theorem is still valid even if the 'external' equilateral triangles are replaced by 'internal' equilateral triangles.

(b) Show that the internal and external Napoleon triangles have the same centroid.

13. (a) On the sides of an arbitrary triangle, construct equilateral triangles so that two of them lie externally while the remaining one lies internally. Show that the centroid of the internal equilateral triangle and the two new vertices of the external equilateral triangles form an isosceles triangle with one of the angles equal to $\frac{2\pi}{3}$.

(b) Show that the conclusion remains valid even if we interchange the roles of 'internal(ly)' and 'external(ly)'.

14. (a) On the sides of an arbitrary triangle ABC, construct equilateral triangles LBC, MCA, NAB externally. Show that LA, MB, NC are the same length, meet at a point, and that the angles of intersection are $\frac{2\pi}{3}$.

(b) What if the equilateral triangles are constructed internally?

15. (a) Construct a triangle, given the three points which are the centers of squares drawn externally on the sides of the desired triangle.

(b) Construct a triangle, given the three points which are the new vertices of equilateral triangles drawn externally on the sides of the desired triangle.

16. (a) Construct similar triangles $P_1A_3A_2$, $A_3P_2A_1$, $A_2A_1P_3$ on the sides of a given $\triangle A_1A_2A_3$, and let G_1, G_2, G_3 be the centroids of those three similar triangles. Show that $\triangle G_1G_2G_3$ is a fourth similar triangle. This generalizes the Napoleon theorem. Actually, we can generalize further by replacing 'centroids' with 'points with the same barycentric coordinates in the respective similar triangles'. (Note, from the order in which we named the vertices of the similar triangles, that the vertices P_1, P_2, P_3 are *not* corresponding vertices of these triangles.)

(b) Let $\triangle Q_1A_3A_2$, $\triangle A_3Q_2A_1$, $\triangle A_2A_1Q_3$ be the reflections of the three similar triangles in (a) with respect to the sides A_2A_3,

A_3A_1, A_1A_2, respectively, and G'_1, G'_2, G'_3 the centroids of those similar triangles. Show that the centroids of $\triangle G_1G_2G_3$ and $\triangle G'_1G'_2G'_3$ coincide if and only if $\triangle P_1A_3A_2$, $\triangle Q_1A_3A_2$, etc., are equilateral.

17. Fix one vertex at $(0, a)$ in the (x, y)-plane, and move another vertex along the x-axis; show that the locus of the third vertex (x, y) of an equilateral triangle consists of two straight lines $y + a = \pm\sqrt{3}x$.

18. Given two lines ℓ and m, and a point A not on ℓ nor on m, construct an equilateral triangle ABC so that the vertex B is on the line ℓ, while the vertex C is on the line m. How many solutions are there?

19. One of the following two 'solutions' to the example at the end of §2.1 is wrong. Which one? State your reason.

    (a) **Trigonometric Solution.** Introduce the notations as in Figure 2.28. Then

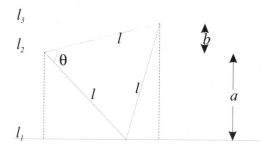

FIGURE 2.28

$$a = \ell \sin\theta,$$

$$b = \ell \sin\left(\frac{\pi}{3} - \theta\right) = \ell\left(\frac{\sqrt{3}}{2}\cos\theta - \frac{1}{2}\sin\theta\right)$$

$$= \frac{\sqrt{3}}{2}\ell\cos\theta - \frac{a}{2}.$$

$$\therefore \ell \cos \theta = \frac{1}{\sqrt{3}}(a + 2b).$$

$$\therefore \ell^2 = a^2 + \frac{1}{3}(a + 2b)^2 = \frac{4}{3}(a^2 + ab + b^2).$$

$$\text{Area} = \frac{\sqrt{3}}{3}(a^2 + ab + b^2).$$

(b) **Slick Solution.** Suppose $f(a, b)$ gives the desired area. Consideration of dimensions allows us to write

$$f(a, b) = pa^2 + qab + rb^2,$$

where p, q, r are coefficients to be determined. If ℓ_2 and ℓ_3 coincide, then $b = 0$ and we have

$$f(a, 0) = \frac{a^2}{\sqrt{3}}. \qquad \therefore p = \frac{1}{\sqrt{3}}.$$

By symmetry, we have $r = p = \frac{1}{\sqrt{3}}$.
If $a = b$, then

$$f(a, a) = \sqrt{3}a^2 = \left(\frac{2}{\sqrt{3}} + q\right) a^2. \qquad \therefore q = \frac{1}{\sqrt{3}}.$$

Therefore,

$$f(a, b) = \frac{1}{\sqrt{3}}(a^2 + ab + b^2).$$

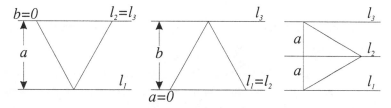

FIGURE 2.29

20. (a) Given three concentric circles, construct an equilateral triangle having one vertex on each of the three given concentric circles. How many solutions are there?

(b) Find the areas of these equilateral triangles in terms of the radii of the concentric circles.

Hint: Let three concentric circles be $|z| = a$, $|z| = b$, $|z| = c$. Fix a vertex at $z = a$, and let another vertex be $z = be^{i\theta}$. Then the third vertex z 'must' satisfy the equation

$$z + be^{i\theta}\omega + a\omega^2 = 0; \quad \text{i.e.,} \quad z = -\omega(a\omega + be^{i\theta}).$$

If $|z| = c$, this gives $a^2 - 2ab\cos\left(\theta+\frac{\pi}{3}\right) + b^2 = c^2$. This means that if we construct a triangle with three sides a, b, c, then the angle opposite the side with length c is $\theta + \frac{\pi}{3}$. This leads us to the following construction. (A reader should carry out the detail and discuss when the construction is possible.)

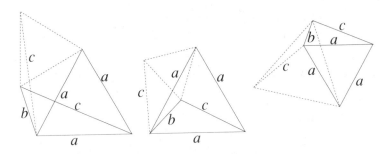

FIGURE 2.30

21. Derive the *law of cosines* from the Ptolemy theorem 2.2.1 by inscribing a trapezoid in a circle.

22. (a) Let $\triangle ABC$ be an equilateral triangle. For any point P on the circumcircle, show that the length of the longest segment among PA, PB, PC is equal to the sum of the lengths of the remaining two shorter ones.

(b) Suppose a point P is on $\overset{\frown}{AD}$ of the circumcircle of a square $ABCD$, show that

$$\overline{PB}\left(\overline{PB} + \overline{PD}\right) = \overline{PC}\left(\overline{PA} + \overline{PC}\right).$$

(c) Let $ABCDE$ be a regular pentagon. For any point P on the circumcircle, show that the sum of the lengths of the longest segment and two shortest ones among PA, PB, PC, PD, PE is equal to the sum of the lengths of the remaining two.

23. Given four points A, B, C, D on a circle, show that the two feet of the perpendiculars from A, B to the line CD, and the two feet of the perpendiculars from C, D to the line AB are cocyclic.

24. For any $a \neq 0$, show that

$$a,\ -\bar{a},\ \frac{1}{a},\ -\frac{1}{\bar{a}},\ 1,\ -1$$

are cocyclic.

25. (a) Given four points A, B, C, D, let A', B', C', D' denote the centroids of the triangles BCD, ACD, ABD, ABC, respectively. Show that A', B', C', D' are cocyclic if and only if A, B, C, D are cocyclic.

(b) What if 'centroids' is replaced by 'orthocenters'? By 'centers of the nine-point circles'?

26. (a) Given $\triangle ABC$ and points P, Q, R on (the extensions of) the sides BC, CA, AB, respectively, show that the circumcircles of $\triangle AQR$, $\triangle BRP$, $\triangle CPQ$ meet at a point.

(b) Given a quadrangle $ABCD$ and points P, Q, R, S on (the extensions of) the sides AB, BC, CD, DA, so that these four points are cocyclic, show that the four new intersections of the neighboring circumcircles of $\triangle APS$, $\triangle BQP$, $\triangle CRQ$, $\triangle DSR$ are cocyclic.

27. (a) Give an example of four circles, any three of which meet at a point without all four of them meeting at a point.

(b) Show that if any three of five circles meet at a point, then all five of them meet at a point.

28. Let D be an arbitrary point on the circumcircle of $\triangle ABC$ and A', B', C' the second intersection of the circumcircle and the perpendiculars from the point D to the sides BC, CA, AB, respectively. Show that AA', BB', CC' are parallel to the Simson line of the point D with respect to $\triangle ABC$.

29. Verify that the equation of the Simson line is self-conjugate.

30. (J. Steiner) Prove that the Simson line of a point on the circumcircle of $\triangle ABC$ bisects the line segment joining that point and the orthocenter of $\triangle ABC$. Prove also that this intersection is on the nine-point circle of $\triangle ABC$.

31. Let D be an arbitrary point on the circumcircle of $\triangle ABC$, and P, Q, R the feet of the perpendiculars from D to the sides BC, CA, AB, respectively. Show that

$$\frac{\overline{AD} \cdot \overline{BC}}{\overline{QR}} = \frac{\overline{BD} \cdot \overline{CA}}{\overline{RP}} = \frac{\overline{CD} \cdot \overline{AB}}{\overline{PQ}}$$

$$= \text{ the diameter of the circumcircle.}$$

32. Given a quadrangle inscribed in a circle, show that the four Simson lines of a vertex with respect to the triangle formed by the remaining three vertices meet at one point, the center of the nine-point circle of the quadrangle.

33. Show that the Simson lines of any two diametrically opposite points on the circumcircle of $\triangle ABC$ are perpendicular to each other and meet on the nine-point circle of $\triangle ABC$.

34. Let D be an arbitrary point on the circumcircle of $\triangle ABC$, and EF a chord perpendicular to the Simson line of D with respect to $\triangle ABC$. Show that the Simson lines of the points D, E, F with respect to $\triangle ABC$ meet at a point.

35. Given $\triangle ABC$ and a point D, let L, M, N be the reflections of the point D with respect to the sides BC, CA, AB, respectively. Show

that the points L, M, N are collinear if and only if the point D is on the circumcircle of $\triangle ABC$; and in this case the line passing through the points L, M, N also passes through the orthocenter H of $\triangle ABC$, and is parallel to the Simson line of D with respect to $\triangle ABC$.

Hint: $\triangle LBC \sim \triangle DBC$ (reversed).

36. (M. B. Cantor) Given n points on a circle, from the centroids of $n - 2$ of these n points drop the perpendiculars to the lines joining the remaining two points. Show that these $\binom{n}{2}$ perpendiculars meet at a point.

37. Given $\triangle ABC$ with $\angle ABC = \frac{2\pi}{3}$, let A', B', C' be the intersections of the angle bisectors at the vertices A, B, C with the respective opposite sides. Show that $\angle A'B'C' = \frac{\pi}{2}$.

38. Let R and r be the circumradius and the in-radius of $\triangle ABC$, and d the distance between the centers of these two circles. Show that

$$d^2 = R(R - 2r).$$

Consequently, $R \geq 2r$.

39. Find the smallest group containing the transformations T_b, T_c and S in the proof of the Morley theorem 2.9.1. How many elements are there in the group?

40. (a) (B. Pascal) Let A, B, C, D, E, F be six arbitrary points on a circle, and P, Q, R be the intersections of (the extensions of) the 'opposite sides' AB and DE, BC and EF, CD and FA, respectively. Show that the points P, Q, R are collinear. This line is called the *Pascal line* of the six points A, B, C, D, E, F.

   (b) Since these points may be labelled in any order, by changing the labelling, we obtain a different Pascal line (for the same six points). How many Pascal lines are possible for the same six points on a circle?

41. (a) Let $\triangle ABC$ be an arbitrary triangle, and P, Q, R be the intersections of the tangent lines to the circumcircle at the vertices

with the extensions of the respective opposite sides. Show that the points P, Q, R are collinear.

(b) Let $ABCD$ be a quadrangle inscribed in a circle, P, Q, the intersections of (the extensions of) AB and CD, AC and BD, respectively, and R, S the intersections of the tangent lines to the circumcircle at A and at D, and at B and at C, respectively. Show that the points P, Q, R, S are collinear.

(c) Given a point on a circle, draw the line tangent to the circle at the given point using only a straightedge.

CHAPTER **3**

Möbius Transformations

3.1 Stereographic Projection

So far complex numbers have been represented by points on a plane. It is often useful to consider them as points on a sphere as well.

Consider a sphere of unit diameter tangent to the complex plane at the origin. In terms of rectangular coordinates (ξ, η, ζ), the equation of

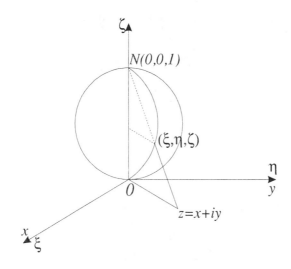

FIGURE 3.1

the sphere is

$$\xi^2 + \eta^2 + \left(\zeta - \frac{1}{2}\right)^2 = \left(\frac{1}{2}\right)^2.$$

For each point $z = x + iy$ on the complex plane, the line segment joining the point to the north pole $N(0, 0, 1)$ meets the sphere at a unique point (other than the north pole). Conversely, for any point on the sphere other than the north pole, the extension of the line segment joining the point to the north pole meets the complex plane at a unique point. Thus we have a one-to-one correspondence between the complex plane and the *Riemann sphere* with the north pole deleted. To remove this exception, we add an ideal point, called the *point at infinity* (denoted ∞), to the complex plane $\mathbb{C}$, and let it correspond to the north pole N. This extended complex plane will be denoted by $\hat{\mathbb{C}}$; thus, $\hat{\mathbb{C}} = \mathbb{C} \cup \{\infty\}$.

Suppose $z = x + iy \in \mathbb{C}$ corresponds to the point (ξ, η, ζ) on the Riemann sphere, then, considering similar triangles, we get

$$\frac{x}{\xi} = \frac{y}{\eta} = \frac{1}{1 - \zeta}. \qquad \therefore x = \frac{\xi}{1 - \zeta}, \qquad y = \frac{\eta}{1 - \zeta}.$$

That is,

$$z = \frac{\xi + i\eta}{1 - \zeta}, \qquad x^2 + y^2 = \frac{\zeta}{1 - \zeta}.$$

Conversely, solving ζ, ξ, η in terms of x, y, and z, we get

$$\xi = \frac{x}{1 + |z|^2} = \frac{z + \bar{z}}{2(1 + |z|^2)},$$

$$\eta = \frac{y}{1 + |z|^2} = \frac{z - \bar{z}}{2i(1 + |z|^2)},$$

$$\zeta = \frac{|z|^2}{1 + |z|^2}.$$

To find the image of a circle (or a line) on the plane to the Riemann sphere, we substitute these relations into the equation of a circle (a line if $A = 0$) in the plane

$$A(x^2 + y^2) + Bx + Cy + D = 0,$$

(where $A, B, C, D \in \mathbb{R}$ and $B^2 + C^2 \geq 4AD$). We get a linear equation

$$A\zeta + B\xi + C\eta + D(1 - \zeta) = 0.$$

Note that the condition that this plane actually intersects the Riemann sphere is:

$$\left| \frac{\frac{1}{2}(A - D) + D}{\sqrt{B^2 + C^2 + (A - D)^2}} \right| \leq \frac{1}{2},$$

which is precisely the condition $B^2 + C^2 \geq 4AD$ that the original equation on the complex plane actually gives a circle. Moreover, if $A = 0$, then the north pole $N(0, 0, 1)$ satisfies the equation of a plane. Since the intersection of a plane and a sphere is a circle, we obtain the first half of the following

THEOREM 3.1.1. *Circles and straight lines on the plane are mapped to circles on the sphere. Straight lines are mapped to circles through the north pole. Conversely, circles on the sphere are mapped into circles and straight lines in the plane.*

Proof. To prove the converse, we observe that a circle on the sphere is the intersection of the sphere with a plane

$$A\xi + B\eta + C\zeta + D = 0,$$

satisfying

$$\left| \frac{\frac{1}{2}C + D}{\sqrt{A^2 + B^2 + C^2}} \right| \leq \frac{1}{2}; \quad \text{i.e.,} \quad A^2 + B^2 \geq 4D(C + D)$$

to ensure actual intersection. In terms of x, y, this equation takes the form

$$(C + D)(x^2 + y^2) + Ax + By + D = 0.$$

If $C + D \neq 0$, then the equation represents a circle; and if $C + D = 0$ (i.e., when the circle on the sphere passes through the north pole), then the equation represents a straight line. $\square$

In particular, this justifies that *any two lines on the plane intersect at the point at infinity*. Another important property of the stereographic projection is the following.

THEOREM 3.1.2. *The stereographic projection is angle-preserving.*

Proof. Without loss of generality, we may assume that two intersecting curves in the plane are straight lines. Now two straight lines in the plane intersecting at the point (x_0, y_0) map into two circles on the sphere intersecting at the point (ξ_0, η_0, ζ_0) and the north pole, and these circles make the same angle with each other at their two intersections. If two lines on the plane are

$$A_1 x + B_1 y + C_1 = 0, \qquad A_2 x + B_2 y + C_2 = 0,$$

then their stereographic images are on the planes

$$A_1 \xi + B_1 \eta + C_1(1 - \zeta) = 0, \qquad A_2 \xi + B_2 \eta + C_2(1 - \zeta) = 0,$$

respectively. The tangents to the corresponding circles at the north pole are the intersections of these planes with the plane $\zeta = 1$; i.e., their equations are

$$A_1 \xi + B_1 \eta = 0, \quad \zeta = 1; \qquad A_2 \xi + B_2 \eta = 0, \quad \zeta = 1,$$

respectively. It is obvious that the angle between the two lines on the complex plane is the same as the angle between two tangent lines at the north pole (since the plane $\zeta = 1$ is parallel to the complex plane). $\square$

Note that, strictly speaking, we need to prove that a curve's property of having a tangent is preserved under the stereographic projection, but that would involve calculus.

3.2 Möbius Transformations

We now discuss elementary but useful functions known as *Möbius transformations* (*linear fractional transformations, bilinear transformations, homographic transformations*, etc.), but first an observation.

To study the behaviors of a real-valued function of a real variable, we use the (x, y)-plane to draw its graph (x for the variable and y for the function)—it enhances our visual understanding and intuition, but this can not be done with complex-valued functions of a complex variable—we would need a four-dimensional space (two dimensions for the variable z and two dimensions for the function w), which is beyond our physical world. So, to study a complex-valued function of a complex variable, we use, instead, two sheets of complex planes, the z-plane for the variable z, and the w-plane for the function w. This is not as ideal as in the case of a real-valued function of a real variable, but this is the best we can manage.

Möbius transformations are defined by special rational functions of the form

$$w = Tz = \frac{\alpha z + \beta}{\gamma z + \delta}, \qquad \begin{vmatrix} \alpha & \beta \\ \gamma & \delta \end{vmatrix} = \alpha\delta - \beta\gamma \neq 0.$$

The condition $\alpha\delta - \beta\gamma \neq 0$ ensures that T is not a constant. Möbius transformations are defined everywhere in the z-plane except at $z = -\frac{\delta}{\gamma}$, and its inverse transformation

$$z = T^{-1}w = \frac{\delta w - \beta}{-\gamma w + \alpha}$$

$\left(\text{note that } \begin{vmatrix} \delta & -\beta \\ -\gamma & \alpha \end{vmatrix} = \begin{vmatrix} \alpha & \beta \\ \gamma & \delta \end{vmatrix} \neq 0\right)$ is also a Möbius transformation and it indicates that every point in the w-plane has a (unique) pre-image except $w = \frac{\alpha}{\gamma}$. It is advantageous to eliminate these exceptions by considering Möbius transformations as mappings from the Riemann sphere onto itself by defining the image of $z = -\frac{\delta}{\gamma}$ to be $w = \infty$, and the pre-image of $w = \frac{\alpha}{\gamma}$ to be $z = \infty$. Noting that the inverse of a Möbius transformation is single-valued, this extension of the definition of Möbius transformations allows us to conclude that Möbius transformations are bijective (one-to-one, onto) mappings of the extended complex plane $\hat{C}$ onto itself.

The identity mapping is clearly a Möbius transformation ($\beta = \gamma = 0$, $\alpha = \delta$). Moreover, a composition of Möbius transformations is a Möbius transformation; i.e., if we map the z-plane onto the w_1-plane by

$$w_1 = Tz = \frac{\alpha z + \beta}{\gamma z + \delta}, \qquad \begin{vmatrix} \alpha & \beta \\ \gamma & \delta \end{vmatrix} \neq 0,$$

and map the w_1-plane onto the w-plane by

$$w = Sw_1 = \frac{aw_1 + b}{cw_1 + d}, \qquad \begin{vmatrix} a & b \\ c & d \end{vmatrix} \neq 0,$$

then the result is a Möbius transformation of the z-plane onto the w-plane given by

$$w = STz = \frac{(a\alpha + b\gamma)z + (a\beta + b\delta)}{(c\alpha + d\gamma)z + (c\beta + d\delta)},$$

with

$$\begin{vmatrix} a\alpha + b\gamma & a\beta + b\delta \\ c\alpha + d\gamma & c\beta + d\delta \end{vmatrix} = \begin{vmatrix} a & b \\ c & d \end{vmatrix} \cdot \begin{vmatrix} \alpha & \beta \\ \gamma & \delta \end{vmatrix} \neq 0.$$

Similarly, the associativity can be verified. We have established the following.

THEOREM 3.2.1. *The set $\mathcal{M}$ of all Möbius transformations forms a **group**; namely, the following four postulates are satisfied.*

(a) *For any two $T, S \in \mathcal{M}$, their **product** (composition) TS is defined, and is an element of $\mathcal{M}$; i.e., $TS \in \mathcal{M}$.*

(b) *$\mathcal{M}$ has an element I, called the **identity**, with the property that*

$$TI = IT = T \qquad \text{for every} \quad T \in \mathcal{M}.$$

(c) *For every element $T \in \mathcal{M}$, there corresponds an element $T^{-1} \in \mathcal{M}$, called the **inverse** of T, with the property that*

$$TT^{-1} = T^{-1}T = I.$$

(d) *The associative law holds in $\mathcal{M}$; for any $T, S, U \in \mathcal{M}$,*

$$(TS)U = T(SU).$$

The expression for a product of Möbius transformations reminds us of matrices. If we write the Möbius transformations T and S above as

$$T = \left(\begin{array}{cc} \alpha & \beta \\ \gamma & \delta \end{array} \right), \quad S = \left(\begin{array}{cc} a & b \\ c & d \end{array} \right),$$

then

$$ST = \left(\begin{array}{cc} a\alpha + b\gamma & a\beta + b\delta \\ c\alpha + d\gamma & c\beta + d\delta \end{array} \right);$$

viz., this notation coincides with the matrix operation, or rather a group of all 2×2 invertible matrices is *homomorphic* to the group of all Möbius transformations.

A subset of a group is called a *subgroup* if it is a group in itself (under the same group operation). Here are some examples of subgroups of the Möbius group $\mathcal{M}$:

EXAMPLE. The set of all *translations*

$$w = Tz = z + b \quad \text{(for some constant } b \in \mathbb{C}\text{)}.$$

EXAMPLE. The set of all *dilations*

$$w = Tz = \alpha z \quad \text{(for some nonzero constant } \alpha \in \mathbb{C}\text{)}.$$

Actually, this subgroup contains two important subgroups:

(a) *Magnifications*:

$$w = Tz = az \quad \text{(where } a \text{ is a positive constant)}.$$

This is just magnifying (or contracting) by the factor a.

(b) *Rotations*:

$$w = Tz = kz \quad \text{(where } |k| = 1\text{)}.$$

This is just a rotation around the origin by the argument of k. Note that magnifications and rotations are commutative and every dilation is a product of a magnification and a rotation.

EXAMPLE.   The subset of $\mathcal{M}$ consisting of the identity and the *reciprocation*:

$$w = Tz = \frac{1}{z}$$

(only) is also a subgroup.

Now let us decompose the rational function defining the Möbius transformation

$$w = Tz = \frac{\alpha z + \beta}{\gamma z + \delta}, \qquad \begin{vmatrix} \alpha & \beta \\ \gamma & \delta \end{vmatrix} \neq 0$$

into partial fractions. There are two possibilities:

(a) If $\gamma = 0$, then $\delta \neq 0$, and we have

$$w = \frac{\alpha}{\delta} z + \frac{\beta}{\delta}.$$

(b) If $\gamma \neq 0$, then

$$w = \frac{\alpha z + \beta}{\gamma z + \delta} = \frac{\alpha}{\gamma} - \frac{(\alpha\delta - \beta\gamma)/\gamma}{\gamma z + \delta}.$$

It follows that a Möbius transformation is a product of dilations, translations and a reciprocation.

It is obvious that translations and dilations map lines to lines and circles to circles. This is not true for the reciprocation. However, it does keep the family of all straight lines *and* circles invariant. Suppose

$$A(x^2 + y^2) + Bx + Cy + D = 0 \qquad (B^2 + C^2 > 4AD),$$

or rather

$$az\bar{z} + \bar{b}z + b\bar{z} + c = 0 \qquad (|b|^2 > ac),$$

where $a = A$ and $c = D$ are real, and $b = \frac{1}{2}(B + iC)$ complex, is an arbitrary circle (line if $a = 0$) in the plane. Performing the reciprocation $w = \frac{1}{z}$, we get

$$a + \bar{b}\bar{w} + bw + cw\bar{w} = 0,$$

which is an equation of the same type. We obtained the following.

THEOREM 3.2.2. *Möbius transformations map the family of all circles and straight lines in the plane to itself.*

As in Chapter 2, *straight lines will be considered as particular cases of circles.*

3.3 Cross Ratios

A Möbius transformation is completely determined by the ratios among its coefficients, hence given three conditions we might be able to find a Möbius transformation that satisfies these conditions. In particular, a circle is determined by three points, thus it might be possible to find a Möbius transformation that maps a given circle in the z-plane onto a given circle in the w-plane.

Let us start from the following observation. Suppose

$$\frac{aw + b}{cw + d} = \frac{\alpha z + \beta}{\gamma z + \delta}.$$

Solving w in terms of z, we obtain w as a Möbius transformation of z; moreover, if the numerator on one side vanishes, then the numerator on the other side must also vanish, and the denominators are related analogously. Therefore, if a Möbius transformation maps z_1, z_2, z_3 to w_1, w_2, w_3, respectively, then we can write

$$\frac{w - w_2}{w - w_3} = k\frac{z - z_2}{z - z_3},$$

where the constant k is to be determined. Regardless of the value of k, the Möbius transformation determined by this equality maps z_2 and z_3 to w_2 and w_3, respectively. Thus, it remains to choose k in such a way that z_1 corresponds to w_1; namely,

$$\frac{w_1 - w_2}{w_1 - w_3} = k\frac{z_1 - z_2}{z_1 - z_3}.$$

Solving for k from the last equality and substituting it into the one above (equivalently, eliminating k from these two equalities by dividing), we get

$$\frac{w - w_2}{w - w_3} \bigg/ \frac{w_1 - w_2}{w_1 - w_3} = \frac{z - z_2}{z - z_3} \bigg/ \frac{z_1 - z_2}{z_1 - z_3}.$$

This gives the Möbius transformation that maps z_1, z_2, z_3 to w_1, w_2, w_3, respectively. Now, if this Möbius transformation also maps z_0 to w_0, then we must have

$$\frac{w_0 - w_2}{w_0 - w_3} \bigg/ \frac{w_1 - w_2}{w_1 - w_3} = \frac{z_0 - z_2}{z_0 - z_3} \bigg/ \frac{z_1 - z_2}{z_1 - z_3},$$

that is,

$$(w_0, w_1; \, w_2, w_3) = (z_0, z_1; \, z_2, z_3).$$

We have established the following.

THEOREM 3.3.1. *The cross ratio is invariant under Möbius transformations.*

COROLLARY 3.3.2. *There exists a Möbius transformation that maps z_0, z_1, z_2, z_3 to w_0, w_1, w_2, w_3, respectively, if and only if*

$$(w_0, w_1; \, w_2, w_3) = (z_0, z_1; \, z_2, z_3).$$

To render this theorem valid in full generality, we extend the definition of the cross ratio for the case when one of the points is ∞ by simply dropping the factors that involve this point. For example, if $z_0 = \infty$, then

$$(z_0, z_1; \, z_2, z_3) := \frac{z_1 - z_3}{z_1 - z_2}.$$

This is quite natural if we observe that

$$(z, z_1; \, z_2, z_3) = \frac{z - z_2}{z - z_3} \bigg/ \frac{z_1 - z_2}{z_1 - z_3}$$

$$= \left(\frac{1 - \frac{z_2}{z}}{1 - \frac{z_3}{z}}\right) \cdot \left(\frac{z_1 - z_3}{z_1 - z_2}\right)$$

$$\longrightarrow \frac{z_1 - z_3}{z_1 - z_2} \qquad (\text{as} \quad z \longrightarrow \infty).$$

It follows that we can always find the Möbius transformation that carries three given points z_1, z_2, $z_3 \in \hat{\mathbb{C}}$ to three preassigned distinct points w_1, w_2, $w_3 \in \hat{\mathbb{C}}$ by simply letting

$$\frac{w - w_2}{w - w_3} \bigg/ \frac{w_1 - w_2}{w_1 - w_3} = \frac{z - z_2}{z - z_3} \bigg/ \frac{z_1 - z_2}{z_1 - z_3},$$

and solving this equation for w. Since the correspondence $z_j \leftrightarrow w_j$ ($j = 1, 2, 3$) completely determines the Möbius transformation, we obtain the desired Möbius transformation.

Since a circle is determined by three points on it, and a Möbius transformation maps a 'circle' to a 'circle', we can thus find Möbius transformations that map a given circle in the z-plane to a given circle in the w-plane. Moreover, three distinct arbitrary points on the first circle can be mapped to three distinct arbitrary points on the second circle.

EXAMPLE. Four points z_0, z_1, z_2, z_3 are cocyclic (or collinear) if and only if their cross ratio $(z_0, z_1; z_2, z_3)$ is real.

Solution. Of course this is Corollary 2.2.2, and we have used it on several occasions already, but here is an alternate solution. Four points are cocyclic (or collinear) if and only if there is a Möbius transformation that maps these points to points on the real axis. Since the cross ratio of four real numbers is real, by the theorem we are done.

THEOREM 3.3.3. *A Möbius transformation is a conformal mapping; i.e., a Möbius transformation preserves the angle (its size as well as its orientation) between two intersecting curves.*

Proof. Without loss of generality, we may assume that the two intersecting curves are circles. Let z_1, z_2 be the intersections of these two circles. Choose arbitrary points z_3 and z_4 on each of the circles, then

$$\arg(z_3, z_4; z_1, z_2) = \arg\left(\frac{z_3 - z_1}{z_3 - z_2}\right) - \arg\left(\frac{z_4 - z_1}{z_4 - z_2}\right)$$

$$= \angle z_2 z_3 z_1 - \angle z_2 z_4 z_1.$$

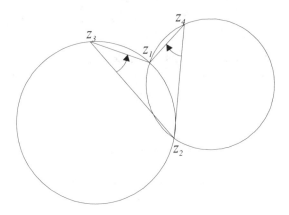

FIGURE 3.2

Let z_3 and z_4 tend to z_1 along the respective circles, then the right-hand side becomes the angle between the two intersecting circles, by Corollary A.2.4. (A reader who is not familiar with the elementary geometry may imitate the argument in the next section.) But the cross ratio is invariant under a Möbius transformation, and hence we obtain the desired result.   □

Suppose a circle C in the z-plane is mapped to a circle C' in the w-plane by a Möbius transformation T. The circle C divides the z-plane into two regions Δ_1 and Δ_2, while the circle C' divides the w-plane into two regions Δ_1' and Δ_2'. We may join any two points z_1 and z_2 in Δ_1 by a circular arc ℓ (or a line segment) not intersecting the circle C. Then the image of ℓ under the Möbius transformation T is a circular arc (or a line segment) that joins the images of z_1 and z_2, and is not intersecting the circle C'. It follows that the images of z_1 and z_2 are either both in Δ_1' or both in Δ_2'. The same holds for two arbitrary points in Δ_2. Since a Möbius transformation is bijective, if for some $z \in \Delta_1$, its image $Tz \in \Delta_1'$, then the image of Δ_1 under T must be the whole Δ_1'; while if $Tz \in \Delta_2'$, then the image of Δ_1 under T must be the whole Δ_2'.

Now consider a circle C and one of its radii. Their images under a Möbius transformation T are a circle C' and a circular arc intersecting at a right angle. Since a Möbius transformation preserves also the orientations of angles, we see immediately that if the interior of the

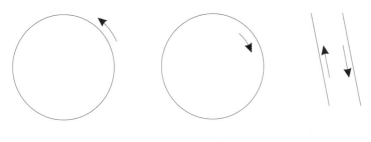

FIGURE 3.3

circle C is mapped to that of the circle C' by a Möbius transformation T, then as a point z moves counterclockwise on the circle C, its image w also moves counterclockwise on the circle C'. On the other hand, if the interior of the circle C is mapped to the exterior of the circle C', then as a point z moves counterclockwise on the circle C, its image w moves clockwise on the circle C'.

Conversely, if a point z and its image w under a Möbius transformation T move on the respective circles with the same orientation, then T maps the interior of the circle C to that of the circle C'; while if z and w move with the opposite orientation, then T maps the interior of C to the exterior of C'.

It is more convenient to adopt the following convention: Consider a circle (or a curve) as an *oriented* curve (usually, by its parametric equation), then as a point moves on the circle (closed curve), the region one sees 'to the left' is the *interior*, by definition. With this convention, a Möbius transformation always maps the interior (of a circle) to the interior (of a circle).

3.4 The Symmetry Principle

Let circles with centers at A and B intersect at a right angle. Suppose a ray from A intersects the circle B at P and Q.

Let T be one of the two intersections of the circles A and B (Our argument works regardless of which one of the two intersections is chosen as T—also only minor modifications are needed if the roles of

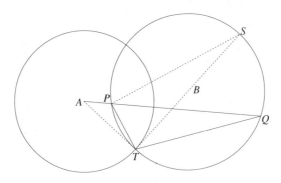

FIGURE 3.4

P and Q are interchanged), and TS a diameter of the circle B. Then

$$\angle AQT = \angle PST = \frac{\pi}{2} - \angle PTS = \angle ATP. \qquad \therefore \ \triangle ATP \sim \triangle AQT.$$

(Note that, in this chapter, we no longer insist that two triangles have the same orientation when we say they are similar.) It follows that

$$\overline{AP} : \overline{AT} = \overline{AT} : \overline{AQ}; \qquad \therefore \ \overline{AP} \cdot \overline{AQ} = \overline{AT}^2 = r^2,$$

where r is the radius of the circle A. In particular, the point Q depends on the circle B only in the sense that the circle B is orthogonal to the circle A and passes through the point P. But there are infinitely many such circles.

Two points P and Q are said to be *symmetric* to (or the *inversion* of) each other with respect to a circle A, if both P and Q are on a ray from (the center of the circle) A *and* $\overline{AP} \cdot \overline{AQ} = r^2$, where r is the radius of the circle A.

The center of the circle and the point at infinity are symmetric to each other, while a point on the circle is symmetric to itself. If a circle degenerates to a line, then two points are *symmetric* to each other if and only if they are reflections of each other with respect to the line.

Our discussion above gives the first half of the following

LEMMA 3.4.1. *If a circle B is orthogonal to a circle A and passes through a point P, then it must also pass through the point Q symmetric to the point*

P with respect to the circle A. Conversely, if a circle B passes through a pair of points P and Q symmetric to each other with respect to a circle A, then the circles A and B are orthogonal to each other.

Proof. The proof of the converse is obtained merely by reversing the argument above. With notations as above, by assumption, we have

$$\overline{AP} \cdot \overline{AQ} = \overline{AT}^2; \quad \text{i.e.,} \quad \overline{AP} : \overline{AT} = \overline{AT} : \overline{AQ}.$$

$$\therefore \ \triangle ATP \sim \triangle AQT.$$

Therefore,
$$\angle ATP = \angle AQT = \angle PST = \frac{\pi}{2} - \angle PTS. \quad \therefore \ \angle ATB = \frac{\pi}{2}. \quad \square$$

THEOREM 3.4.2 (The Symmetry Principle). *A Möbius transformation preserves symmetry.*

Proof. Suppose a pair of points P, Q are symmetric with respect to a circle A, and a Möbius transformation T maps the points P, Q and the circle A to the points P', Q' and the circle A', respectively. We want to show that P', Q' are symmetric with respect to the circle A'.

Let B' be an arbitrary circle passing through the point P' and orthogonal to the circle A'. Then its pre-image $T^{-1}B'$ is a circle orthogonal to the circle A (since T^{-1} is a Möbius transformation, hence is conformal) and passes through the point $T^{-1}P' = P$. By the previous lemma, the circle $T^{-1}B'$ must also pass through the point Q. It follows that the circle B' must pass through the point Q', which implies that Q' is symmetric to P' with respect to A'. $\square$

EXAMPLE. A Möbius transformation that maps the unit circle $|z| = 1$ onto the unit circle $|w| = 1$ must be of the form

$$w = k\frac{z - \alpha}{1 - \bar{\alpha}z}, \quad (|k| = 1, \ |\alpha| \neq 1).$$

Solution. Suppose α $(|\alpha| \neq 1, \ \alpha \neq \infty)$ is the point that is being mapped to $w = 0$, then its symmetric image $\frac{1}{\bar{\alpha}}$ (with respect to the unit circle) must be mapped to $w = \infty$.

$$\therefore \; w = k'\frac{z - \alpha}{z - \frac{1}{\bar{\alpha}}} = k\frac{z - \alpha}{1 - \bar{\alpha}z} \qquad (k = -\bar{\alpha}k').$$

Since $|w| = 1$ when $|z| = 1$, taking $z = 1$, we obtain

$$1 = |k| \cdot \left|\frac{1 - \alpha}{1 - \bar{\alpha}}\right| = |k|.$$

If $\alpha = \infty$, we have $w = \frac{k}{z}$, $|k| = 1$. Conversely, if

$$w = k\frac{z - \alpha}{1 - \bar{\alpha}z}, \qquad (|k| = 1, \; |\alpha| \ne 1),$$

then, for $|z| = 1$,

$$|w| = |k| \cdot \left|\frac{z - \alpha}{1 - \bar{\alpha}z}\right| = \left|\frac{\bar{z}(z - \alpha)}{1 - \bar{\alpha}z}\right| = \left|\frac{1 - \alpha\bar{z}}{1 - \bar{\alpha}z}\right| = 1.$$

Thus the Möbius transformations so obtained satisfy the condition. The interior of the unit circle in the z-plane is mapped to the interior or exterior of the unit circle in the w-plane, depending on whether $|\alpha| < 1$ or $|\alpha| > 1$.

EXAMPLE. A Möbius transformation that maps the real axis in the z-plane to the unit circle $|w| = 1$ in the w-plane must be of the form

$$w = k\frac{z - \mu}{z - \bar{\mu}}, \qquad (|k| = 1, \; \mu \notin \mathbb{R}).$$

Solution. The points on the z-plane corresponding to $w = 0, \infty$ have to be symmetric with respect to the real axis; i.e., they are the complex conjugates of each other. Therefore,

$$w = k\frac{z - \mu}{z - \bar{\mu}}, \qquad (\mu \notin \mathbb{R}).$$

When z is real, $\left|\frac{z - \mu}{z - \bar{\mu}}\right| = 1$, and w has to be on the unit circle; i.e., $|w| = 1$, and so we get $|k| = 1$. Conversely, Möbius transformations

of the above form clearly satisfy the desired condition. The upper half
z-plane is mapped to the interior or exterior of the unit circle $|w| = 1$
in the w-plane depends on whether $\Im\mu > 0$ or $\Im\mu < 0$.

3.5 A Pair of Circles

We have seen already that an arbitrary circle in the z-plane can be
mapped to a preassigned circle in the w-plane by a Möbius transformation. But how about a pair of circles? Can we always map an arbitrary
pair of circles C_1 and C_2 in the z-plane to a preassigned pair of circles C_1'
and C_2' in the w-plane by a Möbius transformation? It is apparent that
if C_1 and C_2 intersect at angle θ, then C_1' and C_2' must also intersect
at angle θ, since a Möbius transformation is conformal. But if this
condition is satisfied, can we guarantee the existence of such a Möbius
transformation?

The answer turns out to be affirmative. To establish this assertion, it is
sufficient to prove that an arbitrary pair of circles C_1 and C_2 intersecting
at angle θ can be mapped by a Möbius transformation to the real axis
and the line $x \sin\theta - y \cos\theta = 0$. (Why is this sufficient?) However,
this is easy to accomplish: Simply map one of the intersections of C_1
and C_2 to the point at infinity, then the images of C_1 and C_2 are two
lines intersecting at angle θ. Now translate the intersection of these two
lines to the origin, rotate by an appropriate angle, and we are done.

In the above argument, we have assumed that $\theta \not\equiv 0 \pmod{\pi}$. (Where
did we use this assumption?) So, what if the pair of circles C_1, and C_2
are tangent to each other? We claim that an arbitrary pair of mutually
tangent circles C_1 and C_2 can be mapped by a Möbius transformation
to a preassigned pair of mutually tangent circles C_1' and C_2'. Again, it
is sufficient to prove that an arbitrary pair of mutually tangent circles
C_1 and C_2 can be mapped by a Möbius transformation to the pair of
parallel lines $y = 0$ and $y = 1$. Again this is easy to accomplish: Simply
map the point of tangency of these two circles to the point at infinity,
then the images of circles C_1 and C_2 become a pair of parallel lines. Now
a translation and a dilation (i.e., a rotation followed by a magnification)
will accomplish our purpose.

It remains to consider the case of a pair of nonintersecting circles.
We claim that a nonintersecting pair of circles can always be mapped by
a Möbius transformation to a pair of concentric circles. To see this, first

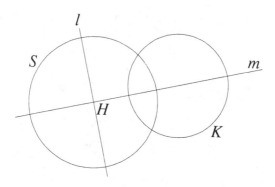

FIGURE 3.5

choose a point on one of the circles, say C_2, and map the point to the point at infinity. Then the images of C_1 and C_2 become a circle and a line not intersecting each other. Call the circle K, and the line ℓ. Let m be the line passing through the center of the circle K and perpendicular to the line ℓ, and H the intersection of the lines ℓ and m. Note that the intersection H is outside the circle K. Draw a circle S with center at H and orthogonal to the circle K. (This can be done by choosing the radius to be the length of the tangent from H to the circle K.) Finally, map one (either one will do) of the intersections of the circle S and the line m to the point at infinity. Then the images of the circle K and the line ℓ are a pair of circles both orthogonal to the images of the circle S and of the line m. But the images of the circle S and the line m are a pair of orthogonal lines. Hence the images of the circle K and the line ℓ must be a pair of concentric circles.

Note carefully: it is *not* true that an arbitrary nonintersecting pair of circles can be mapped to a *preassigned* pair of concentric circles. Given a pair of nonintersecting circles, the ratio of the radii of a pair of concentric circles that the given pair of circles can be mapped to is intrinsic, over which we have no influence.

EXAMPLE.   Let $D = \{z \in \mathbb{C};\ |z| \leq 1\}$ be the closed unit disc, and D' is another closed disc inside D. (In particular, the boundary circles of D and D' do not intersect.) We want to show that there is a Möbius transformation that maps the closed unit disc D onto itself, and maps the disc D' to a disc $\{w \in \mathbb{C};\ |w| \leq r\}$ with a suitable radius r.

Following I. J. Schoenberg (1903–1991) [*Mathematical Time Exposures*, Mathematical Association of America, Washington, D.C., 1982, pp. 188–189], we compute as follows:

By a suitable rotation if necessary, we may assume, without loss of generality, that the center of D' is on the real axis, and

$$[a, b] = D' \cap \{z \in \mathbb{C}; \ \Im z = 0\}$$

is a diameter of D'. If $a + b = 0$, then we have nothing to prove, so without loss of generality, we may assume that $a + b > 0$. (Actually, even if $a + b < 0$, our argument below is valid with only minor modifications.)

Recalling an example in the previous section, and considering that two discs in the z-plane and their images in the w-plane are all symmetric with respect to the real axis, we try a Möbius transformation of the form

$$w = \frac{z - \alpha}{1 - \alpha z},$$

where α is some suitable real number to be determined. Since all the coefficients are real, this Möbius transformation maps the real axis to itself. Moreover, -1 and 1 are two fixed points of this Möbius transformation. Since a Möbius transformation is conformal, and the boundary circles of the discs D and D' intersect the real axis orthogonally, it is sufficient to show that we can find a suitable real number α such that a and b are mapped to $-r$ and r, where r is a real number. This means

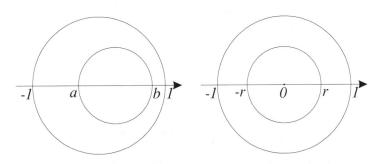

FIGURE 3.6

that

$$\frac{a - \alpha}{1 - \alpha a} + \frac{b - \alpha}{1 - \alpha b} = 0$$

should have real roots. Rewriting the last equation, we get

$$\alpha^2 - \frac{2(1 + ab)}{a + b}\alpha + 1 = 0.$$

Computing ($\frac{1}{4}$ of) its discriminant

$$\left(\frac{1 + ab}{a + b}\right)^2 - 1 = \frac{1 - a^2 - b^2 + a^2b^2}{(a + b)^2}$$

$$= \frac{(1 - a^2)(1 - b^2)}{(a + b)^2} > 0 \qquad (\because -1 < a < b < 1),$$

we see that the equation has real roots. Moreover, from the signs of the coefficients, we see that both roots are positive. Since the product of the two roots is 1 (and $\alpha = 1$ is not a root by our assumption that the boundary circles of D and D' do not intersect), we conclude that one of the roots is between 0 and 1, and the other is greater than 1. Choose the root between 0 and 1 as our α, and we are done.

Remarks. (a) Actually, either choice will do if we merely want to map the two boundary circles to a pair of concentric circles; our choice is influenced by the desire to map the smaller circle to a smaller circle. (b) Even if $D' \supset D$ (i.e., $a < -1$, $b > 1$) our argument is valid with only minor modifications.

THEOREM 3.5.1 (J. Steiner). *Let C, C' be two circles in the plane, one inside the other, say, C' inside C. Draw a circle K_1 tangent to C internally, and to C' externally. Next draw a circle K_2 tangent to C internally, to C' and K_1, externally. Continue this process to get a chain of circles $K_1, K_2, \ldots, K_n$ ($n \geq 3$), where each circle K_j is tangent to C internally, to C', K_{j-1}, K_{j+1} externally ($2 \leq j \leq n - 1$). If it so happens that K_n is tangent to K_1 (and, of course, to C, C' as well as K_{n-1}), then this will also happen regardless of the choice of the position of the initial circle K_1.*

Proof. This is obvious when we map the circles C and C' to two concentric circles via a Möbius transformation. □

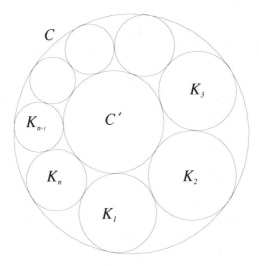

FIGURE 3.7

Remark. Steiner gave a formula that relates the radii r, r' of the circles C, C', the distance d between their centers, and n, for the case when a chain of n circles can be inscribed to be tangent to each other:

$$d^2 = (r - r')^2 - 4rr' \tan^2 \left(\frac{\pi}{n} \right).$$

3.6 Pencils of Circles

Given two circles C_1 and C_2, the set of all circles orthogonal to C_1 and C_2 is called the *pencil of circles* conjugate to C_1 and C_2.

Consider first the case of two intersecting circles C_1 and C_2 in the z-plane. By the discussions in the previous section, we may map C_1 and C_2, by a Möbius transformation to two lines that intersect at the origin in the w-plane. Since a circle is orthogonal to a pair of intersecting lines if and only if its center is at the intersection of the two lines, the pencil of circles conjugate to the pair of circles C_1 and C_2 must be mapped to the set of all concentric circles with the center at $w = 0$. For every point in the extended w-plane, other than the origin and the point at infinity, there is one and only one concentric circle in the pencil that passes through the point. The origin and the point at infinity are called the *limiting points* of the pencil. In terms of the original configuration in the z-plane, this means that exactly one circle in the pencil passes through every point of the extended z-plane, other than the intersections of the circles C_1 and C_2, and no two of these circles in the pencil intersect each other. Pencils of this type are called *hyperbolic* pencils of circles.

We now consider the case of two circles C_1 and C_2 that do not intersect each other. In this case we can map this pair of circles by a Möbius transformation to a pair of concentric circles having the origin of the w-plane as their center. But a 'circle' is orthogonal to a pair of concentric circles if and only if it is a line passing through the center of the concentric circles. Therefore, the image of the pencil of circles conjugate to C_1 and C_2 must be the set of all 'circles' passing through the origin and the point at infinity in the w-plane. In terms of the original configuration in the z-plane, this means that there are two points such that the circles through those two points are the only ones that are orthogonal to the two given circles C_1 and C_2. Those two points are the only pair of points that are symmetric to each other with respect to both C_1 and C_2. They are called the *common points* of the pencil. A pencil of circles consisting of all the circles passing through two fixed points is called an *elliptic* pencil of circles (Figure 3.8).

Finally, we consider the case that two given circles C_1 and C_2 in the z-plane are tangent to each other. In this case we can map C_1 and C_2 to a pair of parallel lines in the w-plane. A 'circle' is orthogonal to a pair of parallel lines if and only if it is a line orthogonal to the pair of parallel lines. So the original pencil of circles conjugate to C_1 and C_2 must be mapped by a Möbius transformation to a set of all lines orthogonal to the pair of parallel lines. In terms of the original configuration in the z-plane, this means that through every point other than the point of tangency of the circles C_1 and C_2, there is one and only one circle passing through the point and it is orthogonal to the circles C_1 and C_2 at their point of tangency. Circles in this pencil have one common point where they are tangent to each other. This type of pencil is called a *parabolic* pencil of circles (Figure 3.8).

Clearly, the type of pencil is invariant under a Möbius transformation. Since the pencil of concentric circles having the origin as their center is orthogonal to the pencil of lines passing through the origin, we see that a hyperbolic pencil of circles may be orthogonal to an elliptic pencil of circles. If two pencils of circles have the property that every circle

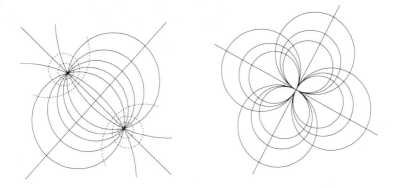

FIGURE 3.8

of one pencil is orthogonal to every circle of the other pencil, then they are called the *conjugate* pencil of circles. Thus every hyperbolic pencil of circles has exactly one conjugate pencil, which turns out to be elliptic; and conversely, every elliptic pencil has exactly one conjugate pencil, which turns out to be hyperbolic. Parabolic pencils can be paired into conjugate pairs. In particular, given two arbitrary circles, there is a unique pencil of circles containing both of them; this pencil is hyperbolic, parabolic, or elliptic, depending on whether the given two circles have 0, 1, or 2 intersections.

3.7 Fixed Points and the Classification of Möbius Transformations

A point z_0 is said to be a *fixed point* of a transformation T if $T z_0 = z_0$. A fixed point of a Möbius transformation

$$w = \frac{az + b}{cz + d}$$

must satisfy the equation

$$cz^2 + (d - a)z - b = 0.$$

Since this is a quadratic equation in z, if a Möbius transformation T has 3 (or more) fixed points, then all the coefficients must vanish; i.e.,

$$c = 0, \qquad d - a = 0, \qquad b = 0,$$

and so T is the identity transformation. In what follows, we shall exclude this case.

(a) If $c = 0$, $a - d = 0$, then T is a translation

$$w = Tz = z + k \qquad \left(k = \frac{b}{a}\right),$$

and so the point at infinity is the unique fixed point of T. If $c = 0$, $a - d \neq 0$, then T is of the form

$$w = Tz = \left(\frac{a}{d}\right) z + \left(\frac{b}{d}\right),$$

and T has two fixed points $\frac{b}{d-a}$ and the point at infinity. In this case, if we set

$$Sz = z - \frac{b}{d-a},$$

then we have

$$S(Tz) = Sw = w - \frac{b}{d-a} = \left(\frac{a}{d}z + \frac{b}{d}\right) - \frac{b}{d-a}$$

$$= \frac{a}{d}\left(z - \frac{b}{d-a}\right) = \frac{a}{d}Sz;$$

that is,

$$T = S^{-1}US,$$

where $Uz = \frac{a}{d}z$ is a dilation.

(b) If $c \neq 0$, $D \neq 0$ (where $D = (d - a)^2 + 4bc$ is the discriminant), then T has two distinct fixed points

$$\alpha = \frac{a - d + \sqrt{D}}{2c}, \quad \text{and} \quad \beta = \frac{a - d - \sqrt{D}}{2c}.$$

In this case, if we set

$$Sz = \frac{z - \alpha}{z - \beta},$$

then we have

$$\frac{w - \alpha}{w - \beta} = k\frac{z - \alpha}{z - \beta};$$

that is,

$$S(Tz) = Sw = k(Sz). \qquad \therefore \ T = S^{-1}US,$$

where

$$Uz = kz \qquad \left(k = \frac{a - \alpha c}{a - \beta c}\right)$$

is a dilation. If $c \neq 0$, $D = 0$, then T has a unique fixed point

$$\alpha = \beta = \frac{a - d}{2c}.$$

Since T maps $z = \alpha$ to $w = \alpha$, we may write

$$\frac{1}{w - \alpha} = \frac{h}{z - \alpha} + k,$$

for suitable constants h and k. Substituting $z = \infty$, $w = \frac{a}{c}$, and $z = 0$, $w = \frac{b}{d}$, we obtain

$$k = \frac{2c}{a + d}, \quad h = 1.$$

Therefore, if we set

$$Sz = \frac{1}{z - \alpha},$$

then

$$S(Tz) = Sw = \frac{1}{w - \alpha} = \frac{1}{z - \alpha} + \frac{2c}{a + d} = V(Sz);$$

that is,

$$T = S^{-1}VS,$$

where $Vz = z + k$ is a translation.

Two Möbius transformations T_1 and T_2 are said to be *similar* if there is a third Möbius transformation S such that $T_2 = S^{-1}T_1S$.

We have established the following.

THEOREM 3.7.1. *Let*

$$w = Tz = \frac{az + b}{cz + d}$$

be a Möbius transformation and $D = (a - d)^2 + 4bc$.

(a) *Case $c = 0$. If $D = 0$, then the point at infinity is the unique fixed point of T, and T can be written in the **normal form**:*

$$w = z + k \qquad \left(k = \frac{b}{d} \right).$$

If $D \neq 0$, then T has two distinct fixed points, $\gamma = \frac{b}{d - a}$ and the point at infinity, and T can be written in the normal form:

$$w - \gamma = k(z - \gamma) \qquad \left(k = \frac{a}{d} \right).$$

(b) *Case $c \neq 0$. If $D \neq 0$, then T has two distinct fixed points*

$$\alpha = \frac{a - d + \sqrt{D}}{2c}, \qquad \beta = \frac{a - d - \sqrt{D}}{2c},$$

and T can be written in the normal form

$$\frac{w - \alpha}{w - \beta} = k \frac{z - \alpha}{z - \beta} \qquad \left(k = \frac{a - \alpha c}{a - \beta c} = \frac{a + d - \sqrt{D}}{a + d + \sqrt{D}} \right).$$

If $D = 0$, then T has a unique fixed point $\alpha = \dfrac{a-d}{2c}$, and T can be written in the normal form

$$\frac{1}{w-\alpha} = \frac{1}{z-\alpha} + k \qquad \left(k = \frac{2c}{a+d} \right).$$

In other words,

(a) a Möbius transformation T has two fixed points if and only if $D \neq 0$, and in this case, T is similar to a dilation; while

(b) T has a unique fixed point if and only if $D = 0$, and in this case, T is similar to a translation.

Remark. Suppose $U_1 z = k_1 z$, $U_2 z = k_2 z$ are two dilations, then U_1 and U_2 are similar if and only if the *multipliers* k_1 and k_2 are either equal to or are the reciprocals of each other; i.e., either $k_1 = k_2$ or $k_1 = \dfrac{1}{k_2}$. However, since a translation $w = z + k$ can be rewritten as $\left(\dfrac{w}{k}\right) = \left(\dfrac{z}{k}\right) + 1$, every translation is similar to the translation $w = z + 1$ (via a dilation); in other words, *any two translations are similar to each other.*

Suppose

$$w = Tz = \frac{az + b}{cz + d}$$

is a Möbius transformation with $D = (d - a)^2 + 4bc \neq 0$. Then there is a Möbius transformation S with the property that

$$Sw = S(Tz) = U(Sz); \qquad \text{i.e.,} \qquad W = UZ,$$

where $W = Sw$, $Z = Sz$, and $UZ = kZ$ is a dilation. If $k > 0$ (but $k \neq 1$), then we say the Möbius transformation T is *hyperbolic*; if $|k| = 1$ (again $k \neq 1$), then we say T is *elliptic*; while all other cases are *loxodromic*.

If T is hyperbolic, then U maps each line passing through the origin in the Z-plane to itself, while mapping a circle with center at the origin to another such circle. In terms of the circles in the original z-plane and w-plane, this means that each circle in the (elliptic) pencil P of circles passing through two fixed points of T is mapped to itself, while every circle in the conjugate (hyperbolic) pencil Q of circles having two fixed points of T as its limiting points is mapped to another circle in the same pencil of circles. In particular, an Apollonius circle

$$\left| \frac{z - \alpha}{z - \beta} \right| = h,$$

where α and β are the fixed points of T, and h is a constant, is mapped to another Apollonius circle

$$\left| \frac{w - \alpha}{w - \beta} \right| = hk.$$

If T is elliptic, then U is a rotation in the Z-plane (by $\arg k$), so we need simply interchange the roles of the pencils P and Q in the preceding paragraph.

If $D = 0$, then we say the Möbius transformation T is *parabolic*, and in this case T has a unique fixed point and is similar to a translation

$$W = VZ = Z + k.$$

Now, V maps every line parallel to the vector k in the Z-plane to itself, while mapping every line perpendicular to the vector k to another line perpendicular to k. In terms of the circles in the original z-plane, this means that there is a (parabolic) pencil P of circles tangent to each other at the unique fixed point of T with the property that every circle in P is mapped to itself by T, while every circle in its conjugate (parabolic) pencil Q (a pencil of circles orthogonal to P at the common point of tangency of circles in P) is mapped to another circle in Q.

3.8 Inversions

Given a point P and a circle with the center at O, we have defined the inversion of the point P with respect to the circle to be the point Q on the ray from the center O through P such that

$$\overline{OP} \cdot \overline{OQ} = r^2,$$

where r is the radius of the circle O. We say O is the *center* of the inversion, and the circle O is the *circle* of inversion.

Most of the results we need about inversions follow from Theorem 3.8.1 and the corresponding results about Möbius transformations.

THEOREM 3.8.1. *Suppose z_j^* are the inversions of z_j ($j = 0, 1, 2, 3$) with respect to some circle. Then*

$$(z_0^*, z_1^*; z_2^*, z_3^*) = \overline{(z_0, z_1; z_2, z_3)}.$$

Proof. Let T be a Möbius transformation that maps the circle of inversion to the real axis. Since a Möbius transformation preserves symmetry, the points z_j and z_j^* are mapped to complex conjugates; i.e., if $w_j = Tz_j$, then $\bar{w}_j = Tz_j^*$ ($j = 0, 1, 2, 3$). Moreover, since a Möbius transformation preserves cross ratios, we have

$$(z_0^*, z_1^*; z_2^*, z_3^*) = (Tz_0^*, Tz_1^*; Tz_2^*, Tz_3^*)$$

$$= (\bar{w}_0, \bar{w}_1; \bar{w}_2, \bar{w}_3) = \overline{(w_0, w_1; w_2, w_3)}$$

$$= \overline{(Tz_0, Tz_1; Tz_2, Tz_3)} = \overline{(z_0, z_1; z_2, z_3)}.$$

$\square$

Since four points are cocyclic if and only if their cross ratio is real (and the complex conjugate of a real number is the real number itself), we immediately obtain the following.

THEOREM 3.8.2. *An inversion preserves circles.*

It is easy to see that an inversion maps the circle of inversion to itself, and a line passing through the center of inversion to itself. Moreover, it maps a circle passing through the center of inversion to a line not passing through the center of inversion, but parallel to the tangent to the circle at the center of inversion. The converse is also true. For, let O be the center of inversion, OA a diameter of the circle being inverted, and B the inversion of A. Then for any point P on the circle, and Q, the inversion of P, we have

$$\overline{OP} \cdot \overline{OQ} = r^2 = \overline{OA} \cdot \overline{OB},$$

(where r is the radius of inversion); i.e.,

$$\overline{OP} : \overline{OA} = \overline{OB} : \overline{OQ}. \qquad \therefore \triangle OAP \sim \triangle OQB.$$

It follows that $\angle OBQ = \angle OPA = \frac{\pi}{2}$.

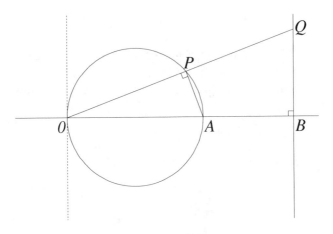

FIGURE 3.9

Conversely, given a line not passing through the center O of the inversion, let B be the foot of the perpendicular from O, and A the inversion of B. Then for any Q on the line, and P the inversion of Q, we have

$$\overline{OP} \cdot \overline{OQ} = \overline{OA} \cdot \overline{OB}; \qquad \overline{OP} : \overline{OA} = \overline{OB} : \overline{OQ}.$$

$$\therefore \triangle OAP \sim \triangle OQB.$$

Therefore, $\angle OPA = \angle OBQ = \frac{\pi}{2}$, and so P is on the circle with OA as its diameter. Finally, a circle not passing through the center of the inversion is mapped to another circle not passing through the center of the inversion.

Imitating our proof of Theorem 3.3.3 that Möbius transformations are conformal (and noting that $\arg\bar{\alpha} \equiv -\arg\alpha$ (mod 2π) for any complex number $\alpha \neq 0$), we obtain the following.

THEOREM 3.8.3. *An inversion is **isogonal**; i.e., an inversion preserves the size of an angle but reverses its orientation.*

Therefore, an inversion maps a pair of intersecting circles to a pair of intersecting circles with the same angle of intersection (neglecting the orientation of the angles). In particular, a pair of orthogonal circles are mapped to a pair of orthogonal circles, and a pair of circles tangent to each other are mapped to a pair of circles tangent to each other. Moreover, a circle orthogonal to the circle of inversion is mapped to itself (by our definition of symmetry).

Our first application of the inversion is the following.

EXAMPLE. (Pappus) Suppose two circles A and B are tangent to each other internally, with the circle B inside the circle A. Let C_0 be the circle having its center on the line AB, and it is tangent to the circle A internally, as well as to the circle B externally. Suppose $C_1, C_2, \ldots$ are circles such that C_1 is tangent to the circles A, B, C_0; C_2 is tangent to the circles A, B, C_1; etc. Then

$$h_n = 2nr_n,$$

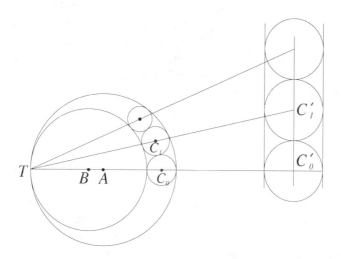

FIGURE 3.10

where h_n is the distance from the center of the circle C_n to the line AB, and r_n is the radius of the circle C_n.

Solution. Let the point T of the tangency of the circles A and B be the center of inversion (with an arbitrary radius of inversion). Then the pair of circles A and B are mapped to a pair of parallel lines perpendicular to the line AB, and the circles $C_0, C_1, C_2, \ldots$ are mapped to circles $C_0', C_1', C_2', \ldots$ tangent to the pair of parallel lines (which are images of the circles A and B), and the radii of the C_n' are all equal (to r', say). Then

$$\frac{r_n}{r'} = \frac{\overline{TC_n}}{\overline{TC_n'}} = \frac{h_n}{\overline{C_n'C_0'}} = \frac{h_n}{2nr'}. \qquad \therefore h_n = 2nr_n.$$

There are many proofs of the Feuerbach theorem 2.8.1 using inversion, but the following proof, which appears in H. S. M. Coxeter and S. L. Greitzer's *Geometry Revisited* [Mathematical Association of America, Washington, D.C., 1967, pp. 117–119], is perhaps the most elementary one known.

THEOREM 3.8.4 (Feuerbach). *The nine-point circle of a triangle is tangent to the incircle and the three excircles.*

Proof. Let I and I_A be the centers of the incircle and the excircle opposite the angle A. Then the three sides of $\triangle ABC$ are three common tangents to the circles I and I_A. Let $B'C'$ be the fourth common tangent where B', C' are on (the extensions of) AC, AB, respectively, and D, E are the feet of perpendiculars from I and I_A to the side BC, respectively. Denote by a, b, c, the lengths of the three sides opposite the angles A, B, C, respectively, and $s = \frac{1}{2}(a + b + c)$. Then it is simple to verify that

$$\overline{BD} = s - b = \overline{CE}.$$

Hence the midpoint L of the side BC is also the midpoint of DE. Therefore, the circle with DE as its diameter is orthogonal to both the incircle I and the excircle I_A. It follows that the inversions of the circles I and I_A with respect to the circle L are circles I and I_A themselves. Since two circles are tangent to each other if and only if their inversions are tangent to each other, it is sufficient to show that the inversion of the nine-point circle is the line $B'C'$.

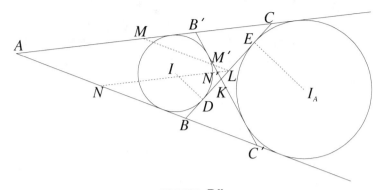

FIGURE 3.11

Since the nine-point circle passes through the center L of the inversion, it will be mapped to a line. Moreover, the nine-point circle passes through the midpoints M, N of the sides CA, AB, so it remains to show that the inversions (with respect to the circle L) of M and N are on the line $B'C'$.

Let K, M', N' be the intersections of $B'C'$ with BC, LM, LN, respectively. Since BI is the bisector of $\angle KBA$, and CI_A is that of the exterior angle of $\angle KCA$, we have, by Lemma A.4.1,

$$\overline{BK} : \overline{AB} = \overline{KI} : \overline{AI} = \overline{CK} : \overline{AC};$$

that is,

$$\frac{\overline{BK}}{c} = \frac{\overline{CK}}{b} = \frac{a}{b+c}. \qquad \therefore \ \overline{BK} = \frac{ac}{b+c}, \quad \overline{CK} = \frac{ab}{b+c}.$$

Since L is the midpoint of BC, this gives

$$\overline{LK} = \frac{a|b-c|}{2(b+c)}.$$

If $b < c$, then simply interchange the roles of B and C (and hence also of b and c), while if $b = c$, the result is trivial, and so we may assume, without loss of generality, that $b > c$. Since $\triangle KLM' \sim \triangle KBC'$, we have

$$\overline{LM'} : \overline{LK} = \overline{BC'} : \overline{BK}; \qquad \text{i.e.,} \qquad \overline{LM'} = \frac{(b-c)^2}{2c},$$

where we have used

$$\overline{BC'} = \overline{AC'} - \overline{AB} = \overline{AC} - \overline{AB} = b - c.$$

$$\therefore \ \overline{LM} \cdot \overline{LM'} = \left(\frac{b-c}{2}\right)^2.$$

Similarly, from $\triangle KLN' \sim \triangle KCB'$, we get

$$\overline{LN'} = \frac{(b-c)^2}{2b}. \qquad \therefore \overline{LN} \cdot \overline{LN'} = \left(\frac{b-c}{2}\right)^2.$$

But

$$\overline{LD} = \frac{1}{2}\overline{DE} = \frac{1}{2}(b-c),$$

and so we are done. □

Before we present another proof of the Ptolemy–Euler theorem 2.2.1, we need a formula describing the effect of inversion on length.

LEMMA 3.8.5. *Suppose P^*, Q^* are the inversions of the points with respect to a circle with center at O and radius r. Then*

$$\overline{P^*Q^*} = \frac{r^2 \cdot \overline{PQ}}{\overline{OP} \cdot \overline{OQ}}, \qquad and \qquad \overline{PQ} = \frac{r^2 \cdot \overline{P^*Q^*}}{\overline{OP^*} \cdot \overline{OQ^*}}.$$

Proof. Since $\triangle OP^*Q^* \sim \triangle OQP$, we have

$$\frac{\overline{P^*Q^*}}{\overline{QP}} = \frac{\overline{OP^*}}{\overline{OQ}} = \frac{r^2}{\overline{OP} \cdot \overline{OQ}}, \qquad \therefore \ \overline{P^*Q^*} = \frac{r^2 \cdot \overline{PQ}}{\overline{OP} \cdot \overline{OQ}}.$$

The second equality is an immediate consequence of the first, since P, Q are the inversions of P^*, Q^*, respectively. □

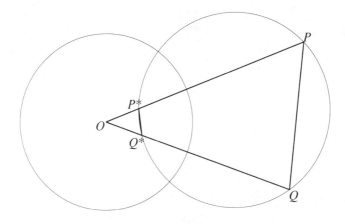

FIGURE 3.12

THEOREM 3.8.6 (Ptolemy–Euler). *For four arbitrary points A, B, C, D,*

$$\overline{AB} \cdot \overline{CD} + \overline{BC} \cdot \overline{DA} \geq \overline{AC} \cdot \overline{BD}.$$

The equality holds if and only if A, B, C, D are cocyclic and in this order.

Proof. Consider an inversion with D as its center. Then, by the triangle inequality, we have

$$\overline{A^*B^*} + \overline{B^*C^*} \geq \overline{A^*C^*},$$

and the equality holds if and only if A^*, B^*, C^* are collinear and in this order. In terms of A, B, C, D, this becomes, by the previous lemma,

$$\frac{r^2 \cdot \overline{AB}}{\overline{AD} \cdot \overline{BD}} + \frac{r^2 \cdot \overline{BC}}{\overline{BD} \cdot \overline{CD}} \geq \frac{r^2 \cdot \overline{AC}}{\overline{AD} \cdot \overline{CD}};$$

that is,

$$\overline{AB} \cdot \overline{CD} + \overline{BC} \cdot \overline{AD} \geq \overline{AC} \cdot \overline{BD},$$

and equality holds if and only if A, B, C, D are cocyclic and are in this order. □

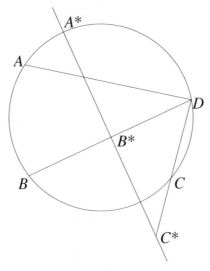

FIGURE 3.13

3.9 The Poincaré Model of a Non-Euclidean Geometry

One of the axioms in Euclidean geometry is that through a point not on a line there exists one and only one line that is parallel to the given line. (This is not the original formulation of Euclid, but is equivalent to it.) Since the parallel axiom is not as self-evident as other axioms, for centuries people tried to prove the parallel axiom from other axioms. It was not until the last century that J. Bolyai (1802–1860) and N. I. Lobachevsky (1793–1856) settled the question by constructing a geometry in which the parallel axiom does not hold. There are two ways to violate the parallel axiom. One way is to construct a geometry in which there are no parallel lines; i.e., every pair of lines intersect. A model for such a geometry is to consider the surface of a sphere where the 'lines' are the great circles. (To be precise, two diametrically opposite points have to be identified. Why?) Another way is to construct a geometry where there is more than one line through a point parallel to a given line. We now present such a model following H. Poincaré (1854–1912).

Our universe or rather 'plane' is the (open) unit disc:

$$U = \{z \in \mathbb{C}; \; |z| < 1\},$$

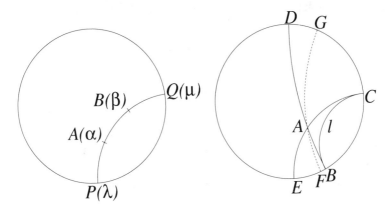

FIGURE 3.14

and 'straight lines' are the circular arcs orthogonal to the unit circle. One can verify that our model satisfies all the axioms in Euclidean geometry except the parallel axiom. For example, given two distinct points A and B in U, there is one and only one 'straight line' that passes through A and B—simply construct a circle that passes through the points A, B, and A^*, the inversion of the point A with respect to the unit circle. (It will automatically pass through B^*, the inversion of the point B.) Now given a 'straight line' ℓ ($=\overset{\frown}{BC}$) and a point A not on ℓ, we can construct 'straight lines' $\overset{\frown}{BAD}$ and $\overset{\frown}{CAE}$ meeting ℓ on the unit circle. Since the points on the unit circle are not in our 'plane' U, 'straight lines' $\overset{\frown}{BAD}$ and $\overset{\frown}{CAE}$ are parallel to the 'straight line' ℓ, and there are infinitely many 'straight lines' such as $\overset{\frown}{FAG}$ 'in between' these two that do not meet ℓ at all.

The 'distance' between two points $z_1, z_2 \in U$ is defined as

$$d(z_1, z_2) := |\log(z_1, z_2; \ \lambda, \mu)|,$$

where λ, μ are the intersections of the circular arc through z_1, z_2 and the unit circle. Note that the distance function $d(z_1, z_2)$ is well-defined: Since z_1, z_2, λ, μ are cocyclic, the cross ratio is a real number, and since z_1, z_2 are not separated by λ and μ, the cross ratio is positive. Moreover,

$d(z_1, z_2)$ is independent of the choice of the order of λ and μ: Since

$$(z_1, z_2; \mu, \lambda) = \frac{1}{(z_1, z_2; \lambda, \mu)},$$

$$|\log(z_1, z_2; \mu, \lambda)| = |-\log(z_1, z_2; \lambda, \mu)|$$

$$= |\log(z_1, z_2; \lambda, \mu)|.$$

Finally, the choice of the base for the logarithm is inessential; it amounts to a change of scale only.

It can be shown that our distance function satisfies the metric axioms:

(a) $d(z_1, z_2) \geq 0$ for all $z_1, z_2 \in U$; equality holds if and only if $z_1 = z_2$.

(b) $d(z_1, z_2) = d(z_2, z_1)$ for all $z_1, z_2 \in U$.

(c) The triangle inequality:

$$d(z_1, z_2) + d(z_2, z_3) \geq d(z_1, z_3) \qquad \text{for all} \quad z_1, z_2, z_3 \in U.$$

Actually, the verification of the first two axioms is a simple muscle exercise.

The concept of angle in the 'plane' U is the usual Euclidean angle between a pair of intersecting circles. But the sum of the (interior) angles of a 'triangle' is less than π.

In Euclidean geometry, we study those properties invariant under rigid motions—the group of transformations consisting of translations, rotations, and reflections. Their counterparts in the Poincaré model are the group(!) of all Möbius transformations that map the unit disc onto itself. Since Möbius transformations preserve cross ratios and are conformal, the distance between two points and the angle between two 'straight lines' are preserved under our 'rigid motions.'

Exercises

1. Suppose z_1, z_2 are points on the complex plane, whose stereographic images on the Riemann sphere are at opposite ends of a diameter. Show that $z_1 \bar{z}_2 = -1$.

2. Show that any four distinct cocyclic points can be mapped to ± 1 and $\pm k$ for some suitable k $(0 < k < 1)$ by a Möbius transformation.

3. Show that the following six Möbius transformations form a group under composition:

$$z, \ \frac{1}{z}, \ 1 - z, \ \frac{1}{1-z}, \ \frac{z}{z-1}, \ \frac{z-1}{z}.$$

4. Show that the cross ratio corresponding to 24 permutations of four points z_0, z_1, z_2, z_3 can have only the following six values:

$$\lambda = (z_0, z_1; z_2, z_3), \ \frac{1}{\lambda}, \ 1 - \lambda, \ \frac{1}{1-\lambda}, \ \frac{\lambda}{\lambda-1}, \ \frac{\lambda-1}{\lambda}.$$

5. Two circles centered at the origin and at the point 1, both with radius 1, together with the line $\Re z = \frac{1}{2}$ divide the complex plane $\mathbb{C}$ into six regions. Show that each of these six regions contains exactly one of the following six values:

$$\lambda, \ \frac{1}{\lambda}, \ 1 - \lambda, \ \frac{1}{1-\lambda}, \ \frac{\lambda}{\lambda-1}, \ \frac{\lambda-1}{\lambda},$$

provided that λ is not on any of the boundaries.

6. Find a nonconstant function $f(z)$ satisfying the functional equation

$$f(z) = f\left(\frac{1}{z}\right) = f(1-z) = f\left(\frac{1}{1-z}\right)$$

$$= f\left(\frac{z}{z-1}\right) = f\left(\frac{z-1}{z}\right) \quad (z \neq 0, 1).$$

How many such functions are there?

7. Find a function $f(z)$ satisfying the functional equation

$$f\left(\frac{1}{1-z}\right) + f\left(\frac{z-1}{z}\right) = \frac{z}{z-1} \quad \text{for all} \quad z \neq 0, 1.$$

Is such a function unique?

8. A point z_0 is said to be a *fixed point* of a transformation T if $Tz_0 = z_0$.

   (a) Show that a Möbius transformation that has $0, 1, \infty$ as its fixed points must be the identity transformation.

   (b) Show that a Möbius transformation that has three distinct fixed points must be the identity transformation.

   (c) Suppose Möbius transformations S and T have the property that $Sz_i = Tz_i$ for three distinct points z_i ($i = 1, 2, 3$), show that $S = T$.

9. Find the Möbius transformation that maps $-1, 0, 1$ to $-i, 1, i$.

10. Check the invariance of cross ratio for translation, dilation, and reciprocation, by computation.

11. Using the property that a Möbius transformation preserves cross ratios, show that a Möbius transformation maps a circle to a circle.

12. Given a pair of intersecting circles, the line passing through the centers of these two circles intersects the circles at four points. Show that one of the cross ratios of these four points is $\sin^2 \frac{\theta}{2}$, where θ is the angle of intersection of the circles.

13. (a) Given a point P outside a circle C, using a straightedge and compass, find the point Q symmetric to the point P with respect to the circle C.

   (b) What if the point P is inside the circle C?

   (c) Given a point P and two circles C_1 and C_2, construct a circle C that passes through the point P and is orthogonal to both C_1 and C_2.

   (d) Given a pair of nonintersecting circles, find the pair of points symmetric to both circles.

14. (a) Show that z_0 and z_0^* are symmetric with respect to the circle

$|z - a| = r$ if and only if

$$(\bar{z}_0 - \bar{a})(z_0^* - a) = r^2.$$

(b) Show that z_0 and z_0^* are symmetric with respect to the circum-circle of $\triangle z_1 z_2 z_3$ if and only if

$$(z_0^*, z_1; z_2, z_3) = \overline{(z_0, z_1; z_2, z_3)}.$$

(c) Use (b) to prove the symmetry principle 3.4.2.

15. (a) If z_0 and z_0^* are symmetric with respect to a circle C, and z_1, z_2 are two arbitrary points on the circle C, show that the cross ratio $(z_0, z_0^*; z_1, z_2)$ has the absolute value 1.

(b) Given three points z_1, z_2, z_3, describe the set of all z_0 satisfying

$$|(z_0, z_1; z_2, z_3)| = 1.$$

16. (a) Show that any Möbius transformation that maps the real axis to itself can be expressed with real coefficients.

(b) Show that any Möbius transformation that maps the real axis in the z-plane to the unit circle in the w-plane can be expressed as

$$w = \frac{\alpha z - \beta}{\bar{\alpha} z - \bar{\beta}} \qquad \left(\frac{\beta}{\alpha} \notin \mathbb{R}\right).$$

(c) Show that any Möbius transformations that maps the unit circle to itself can be expressed as

$$w = \frac{\alpha z - \beta}{\bar{\beta} z - \bar{\alpha}} \qquad (|\alpha| \neq |\beta|).$$

17. (a) Find a Möbius transformation that maps the circles $|z| = 1$ and $|z - \frac{1}{3}| = \frac{1}{2}$ to a pair of concentric circles. What is the ratio of the radii of the concentric circles?

(b) Same problem for $|z| = 1$ and $|z + 2| = 4$.

18. Find a Möbius transformation that maps the real axis and the circle $|z - 5i| = 4$ to two concentric circles with center at the origin. What is the ratio of the radii?

19. Suppose that a Möbius transformation maps a pair of concentric circles to another pair of concentric circles. Show that the ratio of the radii is preserved.

20. (a) Find a Möbius transformation that maps a pair of circles $|z - 1| = 1$ and $|z - 2| = 2$ to a pair of circles $|w - 1| = 1$ and $|w + 1| = 1$.

    (b) Find a Möbius transformation that maps a pair of circles $|z| = 1$ and $|z - \sqrt{3}| = 2$ to a pair of circles $|w - 1| = 2$ and $|w + 1| = 2$.

    (c) Find a Möbius transformation that maps a pair of circles $|z| = 1$ and $|z - 5| = 3$ to a pair of concentric circles.

21. Find the fixed points of the Möbius transformations

    (a) $w = \dfrac{z}{z - 1}$;

    (b) $w = \dfrac{z - 4}{z - 3}$;

    (c) $w = \dfrac{z}{2 - z}$;

    (d) $w = \dfrac{5z - 4}{2z - 1}$.

    Which of these transformations is elliptic? Hyperbolic? Parabolic? Loxodromic?

22. (a) Find all the Möbius transformations T such that T^2 is the identity.

    (b) Show that T^n is the identity if and only if it has the normal form

    $$\frac{w - \alpha}{w - \beta} = k\frac{z - \alpha}{z - \beta} \qquad (k^n = 1).$$

23. (a) Is the reflection $w = Tz = \bar{z}$ with respect to the real axis a Möbius transformation?

(b) How about inversion with respect to the unit circle?

24. Find the images under inversion with respect to the unit circle of

   (a) the line $y = x$;

   (b) the line $x = 1$;

   (c) the circle $|z - 2| = 1$;

   (d) the circle $|z - 2| = 2$.

25. Does an inversion preserve symmetry?

26. Is the composition of two inversions an inversion? Is it a Möbius transformation?

27. Given a line ℓ and a circle C with center at A, let B be the point symmetric to A with respect to the line ℓ, and B^* the inversion of B with respect to the circle C. Show that B^* is the center of the circle which is the inversion of the line ℓ with respect to the circle C.

28. (a) Given a line and a circle, show that there is an inversion that maps one to the other.

   (b) Same problem with a pair of circles instead of a line and a circle.

29. Given a pair of nonintersecting circles C_1 and C_2, show that there is an inversion that maps C_1 and C_2 to a pair of concentric circles.

30. Is there a Möbius transformation that maps the unit disc onto itself, and interchanges two given points in the unit disc? If so, how many such Möbius transformations are there?

31. In the Poincaré model, suppose lines ℓ_1 and ℓ_2 are parallel, as are the lines ℓ_2 and ℓ_3. Is it necessary that ℓ_1 and ℓ_3 are parallel?

32. Show that the distance $d(z_1, z_2)$ in the Poincaré model tends to ∞ as z_2 tends to a point on the unit circle.

33. Suppose z_1, z_2, z_3 are on a 'straight line' in this order. Show that

$$d(z_1, z_2) + d(z_2, z_3) = d(z_1, z_3).$$

34. Suppose z_1, z_2, w_1, w_2 are points in the Poincaré plane. Show that

$$d(z_1, z_2) = d(w_1, w_2)$$

if and only if there is a Möbius transformation mapping the unit disc onto itself such that $T z_j = w_j$ $(j = 1, 2)$.

Hint: First find Möbius transformations that map z_2 and w_2 to the origin.

35. Prove the triangle inequality

$$d(z_1, z_2) + d(z_2, z_3) \geq d(z_1, z_3) \qquad (z_1, z_2, z_3 \in U)$$

for the Poincaré model.

Hint: Use a Möbius transformation to map the point z_2 to the origin.

36. Describe a 'circle' in the Poincaré model.

37. Given a 'straight line' ℓ in the Poincaré model, what is the locus of the point z such that the distance from z to the given 'straight line' ℓ is constant?

38. Show that the set of all Möbius transformations that map the unit disc onto itself forms a group.

Epilogue

I hope I have succeeded in demonstrating that the complex numbers are useful, and that they can be employed to produce easy proofs of many beautiful results in plane geometry.

Since the subject is so old, it is hard to claim that a certain result is new, but to the best of my knowledge, the proofs (as presented here) of Theorems 2.5.1, 2.6.1, 2.6.3, 2.9.1 (and its variations), as well as Theorem 2.6.2 do not seem to have appeared in print. At least *I could swear I did these myself*—to borrow an expression from Y. Katznelson. Naturally, many exercise problems are my own creations. I am particularly fond of Exercises 8 and 16 (b) in Chapter 2.

In writing Chapter 1, I consulted

K. Katayama, *Geometry of Complex Numbers*, Iwanami, Tokyo, 1982 (in Japanese)

frequently, and for Chapter 2,

K. Yano, *Famous Theorems in Geometry*, Kyoritsu, Tokyo, 1981 (in Japanese).

In fact, the latter work inspired me to structure my lectures around complex numbers. I am deeply grateful to these two authors.

Finally, for those readers who wish to pursue the subject further, the book

H. Schwerdtfeger, *Geometry of Complex Numbers*, Dover, New York, 1979

is highly recommended.

Preliminaries in Geometry

A.I Centers of a Triangle

A.1.1   The Centroid.

LEMMA A.1.1. *Let D, E be the midpoints of the sides AB, AC of $\triangle ABC$. Then*

$$DE \parallel BC \qquad and \qquad \overline{DE} = \frac{1}{2}\overline{BC}.$$

Proof. Extend DE to F so that $\overline{DE} = \overline{EF}$. Then in $\triangle ADE$ and $\triangle CFE$,

$$\overline{AE} = \overline{CE}, \qquad \overline{DE} = \overline{FE}, \qquad \angle AED = \angle CEF;$$

$$\therefore \ \triangle ADE \cong \triangle CFE.$$

It follows that

$$\overline{CF} = \overline{AD} = \overline{DB},$$

and

$$\angle CFE = \angle ADE. \qquad \therefore \ CF \parallel BD.$$

Thus the quadrangle $BCFD$ is a parallelogram.

$$\therefore \ \overline{DE} = \frac{1}{2}\overline{DF} = \frac{1}{2}\overline{BC}, \quad \text{and} \quad DE \parallel BC.$$

☐

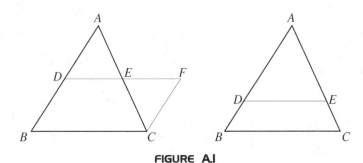

FIGURE A.1

Actually, this lemma is a particular case of the following theorem.

THEOREM A.1.2. *Suppose D, E are points on the sides AB, AC of $\triangle ABC$ such that $DE \parallel BC$. Then*

$$\frac{\overline{AD}}{\overline{AB}} = \frac{\overline{AE}}{\overline{AC}} = \frac{\overline{DE}}{\overline{BC}}.$$

The converse is also true.

Proof. $\triangle ADE \sim \triangle ABC$.   ☐

THEOREM A.1.3. *The three medians of a triangle meet at a point. This point is called the* **centroid** *of the triangle.*

Proof. Let G be the intersection of the medians BD and CE of $\triangle ABC$. Extend AG to F so that $\overline{GF} = \overline{AG}$. Then in $\triangle ABF$, E and G are the midpoints of sides AB and AF, respectively. Hence, by the previous lemma,

$$BF \parallel EG \parallel GC.$$

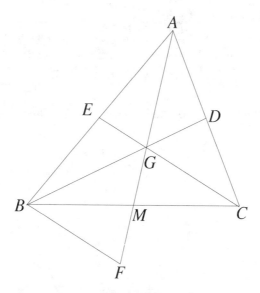

FIGURE A.2

Similarly, $CF \parallel GB$. Therefore, the quadrangle $BFCG$ is a parallelogram, and so its two diagonals bisect each other, say at M. We have shown that the extension of AG passes through the midpoint M of the side BC; i.e., the three medians of a triangle intersect at a point.   □

Note that from our proof,

$$\overline{AG} = \overline{GF} = 2\overline{GM}, \quad \overline{CG} = \overline{FB} = 2\overline{GE},$$

and similarly,

$$\overline{BG} = 2\overline{GD}.$$

A.1.2   The Circumcenter.

LEMMA A.1.4. *Suppose A and B are two fixed points. Then a point P is on the perpendicular bisector of the line segment AB if and only if $\overline{PA} = \overline{PB}$.*

Proof. Suppose P is on the perpendicular bisector of the line segment AB. Join the point P and the midpoint M of AB. Then

$$\triangle PAM \cong \triangle PBM \quad \text{(by SAS)},$$

and so $\overline{PA} = \overline{PB}$.

Conversely, if $\overline{PA} = \overline{PB}$, then

$$\triangle PAM \cong \triangle PBM \quad \text{(by SSS)},$$

where M is the midpoint of the line segment AB. Therefore,

$$\angle AMP = \angle BMP = \frac{\pi}{2}.$$

$\square$

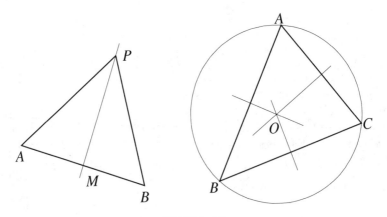

FIGURE A.3

THEOREM A.1.5. *The three perpendicular bisectors of the sides of a triangle meet at a point. This point is called the **circumcenter** of the triangle.*

Proof. Let O be the intersection of the perpendicular bisectors of the sides AB and AC. Then, since the point O is on the perpendicular bisector of AB, by the first part of the lemma, we have $\overline{BO} = \overline{AO}$. Similarly, since O is also on the perpendicular bisector of AC, we have $\overline{AO} = \overline{CO}$. But then $\overline{BO} = \overline{CO}$, and so, by the second half of the lemma, the point O is on the perpendicular bisector of the side BC. $\square$

Note that since the distances from the point O to the three vertices are equal, if we draw a circle with O as the center and $\overline{OA}$ as its radius, we obtain a circumcircle of $\triangle ABC$.

A.1.3 The Orthocenter.

THEOREM A.1.6. *The three perpendiculars from the vertices to the opposite sides of a triangle meet at a point. This point is called the **orthocenter** of the triangle.*

Proof. Through each vertex of $\triangle ABC$, draw a line parallel to the opposite side, obtaining $\triangle A'B'C'$. Then the quadrangles $ABCB'$ and $ACBC'$ are parallelograms, and so $\overline{AB'} = \overline{BC} = \overline{C'A}$. Since $B'C' \parallel BC$, the perpendicular from the vertex A to the side BC is the perpendicular bisector of the line segment $B'C'$. In other words, the three perpendiculars from the vertices of $\triangle ABC$ to the opposite sides are the perpendicular bisectors of the three sides of $\triangle A'B'C'$. Hence, by the previous theorem, these three lines meet at a point. $\qquad \square$

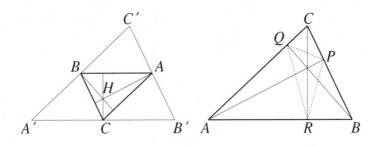

FIGURE A.4

Alternate Proof. Let P, Q, R be the feet of the perpendiculars from the vertices A, B, C to the respective opposite sides BC, CA, AB of $\triangle ABC$. Observe that

$$\angle BQC = \angle BRC \left(= \frac{\pi}{2} \right),$$

and so, by Corollary A.2.2 below, B, R, Q, C are cocyclic. Similarly, C, P, R, A are cocyclic, so are A, Q, P, B. Therefore, by Lemma A.2.1

below,
$$\angle APQ = \angle ABQ = \angle QCR = \angle APR.$$

Similarly,
$$\angle BQR = \angle BQP, \quad \angle CRP = \angle CRQ.$$

We have shown that the three perpendiculars of $\triangle ABC$ are the three angle bisectors of the pedal triangle PQR. Hence, they meet at the incenter of $\triangle PQR$, by Theorem A.1.8 below.   □

A.1.4 The Incenter and the Three Excenters.

LEMMA A.1.7. *Let P be a point inside $\angle BAC$. Then P is on the bisector of $\angle BAC$ if and only if the distances from the point P to the sides AB and AC are equal.*

Proof. Let P be an arbitrary point on the bisector of $\angle BAC$, and D, E the feet of the perpendiculars from P to AB and AC, respectively. Then in $\triangle APD$ and $\triangle APE$, two pairs of corresponding angles are equal and so these two triangles are similar. Moreover, they have a corresponding side AP in common, hence

$$\triangle APD \cong \triangle APE. \qquad \therefore \ \overline{PD} = \overline{PE}.$$

Conversely, suppose P is a point inside $\angle BAC$ such that $\overline{PD} = \overline{PE}$, where D and E are the feet of the perpendiculars from the point P to AB and AC, respectively. Then, by the Pythagorean theorem, three pairs of corresponding sides of $\triangle APD$ and $\triangle APE$ are equal, and so

$$\triangle APD \cong \triangle APE. \qquad \therefore \ \angle PAD = \angle PAE.$$

□

THEOREM A.1.8. *The three bisectors of the (interior) angles of a triangle meet at a point. This point is called the **incenter** of the triangle.*

Proof. Let I be the intersection of the bisectors of the angles at the vertices B and C of $\triangle ABC$, and D, E, F the feet of the perpendiculars from I to the three sides BC, CA, AB, respectively. Then, since I is on

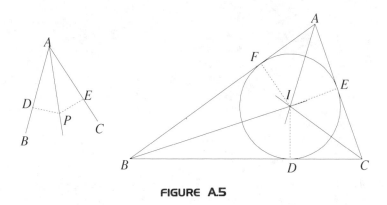

FIGURE A.5

the bisector of $\angle ABC$, by the first part of the lemma, we have $\overline{IF} = \overline{ID}$. Similarly, since I is also on the bisector of $\angle ACB$, we have $\overline{ID} = \overline{IE}$. $\therefore$ $\overline{IE} = \overline{IF}$. But then, by the second part of the lemma, I must be on the bisector of $\angle BAC$. □

Since the distances from the incenter I to the three sides of a triangle are all equal, if we draw a circle with center at I and use the distance from I to a side as the radius, we obtain the circle tangent to all three sides of the triangle. This circle is called the *incircle* of the triangle.

THEOREM A.1.9. *The bisectors of two exterior angles and that of the remaining interior angle of a triangle meet at a point. This point is called an **excenter** of the triangle, and is the center of an **excircle** that is tangent to extensions of two sides and the remaining side of a triangle. A triangle has three excenters and three excircles. (See Figure A.6.)*

Proof. The proof is essentially the same as that for the incenter (and the incircle). □

A.1.5 Theorems of Ceva and Menelaus. Each centroid, orthocenter, incenter, and excenter is the intersection of three lines passing through the vertices of a triangle. For this type of problem, the following theorem of G. Ceva (1647–1734) is very effective.

THEOREM A.1.10. *Let P, Q, R be points on (the extensions of) the respective sides BC, CA, AB of △ABC. Then the lines AP, BQ, CR meet at*

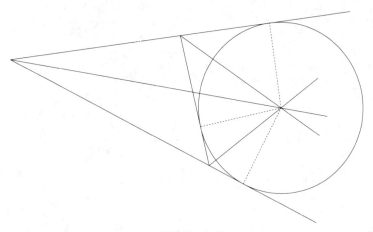

FIGURE A.6

a point if and only if

$$\frac{\overline{BP}}{\overline{PC}} \cdot \frac{\overline{CQ}}{\overline{QA}} \cdot \frac{\overline{AR}}{\overline{RB}} = 1.$$

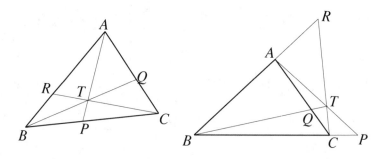

FIGURE A.7

For example, to prove that the three medians of a triangle meet at a point, using previous notation, we have

$$\frac{\overline{BM}}{\overline{MC}} = \frac{\overline{CD}}{\overline{DA}} = \frac{\overline{AE}}{\overline{EB}} = 1,$$

and so the condition in the Ceva theorem is clearly satisfied.

In the case of the three perpendiculars, let

$$a = \overline{BC}, \quad b = \overline{CA}, \quad c = \overline{AB}, \quad \alpha = \angle A, \quad \beta = \angle B, \quad \gamma = \angle C,$$

and let P, Q, R be the feet of the perpendiculars from the vertices A, B, C to the respective opposite sides, then $\overline{BP} = c\cos\beta$, etc., and so

$$\frac{\overline{BP}}{\overline{PC}} \cdot \frac{\overline{CQ}}{\overline{QA}} \cdot \frac{\overline{AR}}{\overline{RA}} = \frac{c\cos\beta}{b\cos\gamma} \cdot \frac{a\cos\gamma}{c\cos\alpha} \cdot \frac{b\cos\alpha}{a\cos\beta} = 1,$$

and we are done.

To prove that the three angle bisectors meet at a point, let U, V, W be the intersections of the angle bisectors at A, B, C and the respective opposite sides. Then, by Lemma A.4.1 to the Apollonius circle below, we have $\frac{\overline{BU}}{\overline{UC}} = \frac{b}{c}$, etc.,

$$\therefore \quad \frac{\overline{BU}}{\overline{UC}} \cdot \frac{\overline{CV}}{\overline{VA}} \cdot \frac{\overline{AW}}{\overline{WB}} = \frac{b}{c} \cdot \frac{c}{a} \cdot \frac{a}{b} = 1.$$

The case of an excenter is essentially the same as that of the incenter, and so is left for the reader.

It remains to prove the Ceva theorem itself. Suppose AP, BQ, CR meet at a point, say T. Draw the line passing through the point A parallel to the side BC meeting (the extensions of) BQ, CR at B', C', respectively. Since

$$\triangle BPT \sim \triangle B'AT, \qquad \triangle CPT \sim \triangle C'AT,$$

we have

$$\frac{\overline{BP}}{\overline{PT}} = \frac{\overline{B'A}}{\overline{AT}}, \quad \frac{\overline{PT}}{\overline{PC}} = \frac{\overline{AT}}{\overline{AC'}}. \qquad \therefore \ \frac{\overline{BP}}{\overline{PC}} = \frac{\overline{B'A}}{\overline{AC'}}.$$

Similarly,

$$\frac{\overline{CQ}}{\overline{QA}} = \frac{\overline{CB}}{\overline{AB'}}, \quad \text{and} \quad \frac{\overline{AR}}{\overline{RB}} = \frac{\overline{AC'}}{\overline{BC}}.$$

Hence multiplying the last three equalities together, we get the desired equality.

To prove the converse, let T be the intersection of (the extensions of) BQ and CR, and P' the intersection of (the extensions of) AT and BC. Then, by what we have shown,

$$\frac{\overline{BP'}}{\overline{P'C}} \cdot \frac{\overline{CQ}}{\overline{QA}} \cdot \frac{\overline{AR}}{\overline{RB}} = 1.$$

On the other hand, by assumption, we also have

$$\frac{\overline{BP}}{\overline{PC}} \cdot \frac{\overline{CQ}}{\overline{QA}} \cdot \frac{\overline{AR}}{\overline{RB}} = 1. \qquad \therefore \ \frac{\overline{BP'}}{\overline{P'C}} = \frac{\overline{BP}}{\overline{PC}}.$$

Now adding 1 to both sides, we get

$$\frac{\overline{BP'} + \overline{P'C}}{\overline{P'C}} = \frac{\overline{BP} + \overline{PC}}{\overline{PC}}. \qquad \therefore \ \frac{\overline{BC}}{\overline{P'C}} = \frac{\overline{BC}}{\overline{PC}}.$$

It follows that $\overline{P'C} = \overline{PC}$, and so $P' = P$.

(We have deliberately suppressed an accompanying configuration, giving a reader a chance to check that the proof works for all cases.)

As the careful reader will notice, the converse holds if and only if the line segments are considered as *directed* : Namely, $\frac{\overline{BP}}{\overline{PC}} > 0$ if BP and PC are in the same direction, and $\frac{\overline{BP}}{\overline{PC}} < 0$ if BP and PC are in the opposite direction. Similar considerations naturally apply to the other ratios.

The following theorem, closely associated with that of Ceva, was discovered by Menelaus of Alexandria (ca. 98):

THEOREM A.1.11. *Points P, Q, R on (the extensions) of the respective sides BC, CA, AB of △ABC are collinear if and only if*

$$\frac{\overline{BP}}{\overline{PC}} \cdot \frac{\overline{CQ}}{\overline{QA}} \cdot \frac{\overline{AR}}{\overline{RB}} = -1.$$

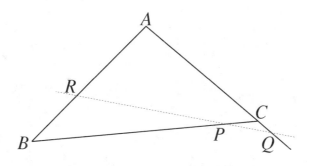

FIGURE A.8

Proof. Since we shall not need this theorem, we merely sketch a proof, and leave the details for the reader. To prove that the condition is necessary, draw a line passing through the vertex A parallel to the line determined by the points P, Q, R, meeting (the extension of) the side BC at A'. Now express all the ratios involved in terms of those of the segments on the line BC. To prove sufficiency, imitate the proof of the Ceva theorem. □

A.2 Angles Subtended by an Arc

LEMMA A.2.1. *An angle subtended by an arc is equal to one half of its central angle. In particular, all the angles subtended by the same arc are equal.*

Proof. Let A, B, C be points on a circle O.

Case 1. Suppose the center O is either on the segment AC or on BC. To fix our notation, let the center O be on the chord AC. Then, since $\triangle OBC$ is an isosceles triangle, we have $\angle OCB = \angle OBC$. But $\angle AOB$ is an exterior angle of $\triangle OBC$,

$$\therefore \ \angle AOB = \angle OBC + \angle OCB = 2\angle ACB.$$

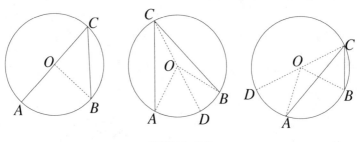

FIGURE A.9

Case 2.   Suppose the center O is inside $\angle ACB$. Let CD be a diameter. Then from Case 1, we have

$$\angle ACB = \angle ACD + \angle DCB$$
$$= \frac{1}{2}(\angle AOD + \angle DOB) = \frac{1}{2}\angle AOB.$$

Case 3.   Suppose the center O is outside $\angle ACB$. As before, let CD be a diameter. Then, again from Case 1, we have

$$\angle ACB = \angle BCD - \angle ACD$$
$$= \frac{1}{2}(\angle BOD - \angle AOD) = \frac{1}{2}\angle AOB.$$

$\square$

THEOREM A.2.2. *Suppose points C and D are on the same side of a line AB. Then the points A, B, C, D are cocyclic if and only if $\angle ACB = \angle ADB$.*

Proof. It remains to prove the converse. Draw the circle passing through the points A, B, and C. Suppose the point D is inside this circle. Let D' be the intersection of the circle and the extension of AD, then

$$\angle ADB = \angle AD'B + \angle DBD' > \angle AD'B = \angle ACB.$$

Now if the point D is outside of this circle, let D' be the intersection of the circle and AD. Then

$$\angle ADB < \angle ADB + \angle DBD' = \angle AD'B = \angle ACB.$$

Hence, if $\angle ADB = \angle ACB$, then the point D must be on the circle passing through the points A, B, C (and in this case, the equality clearly holds, by the previous lemma). □

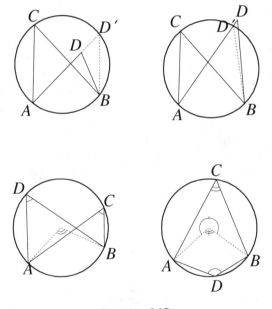

FIGURE A.10

COROLLARY A.2.3. *Suppose C and D are on opposite sides of a line AB. Then the points A, B, C, D are cocyclic if and only if*

$$\angle ACB + \angle ADB = \pi.$$

COROLLARY A.2.4. *The angle between a tangent to a circle and a chord is equal to angles subtended by the arc inside this angle.*

Proof. Pictures are worth a thousand words.   □

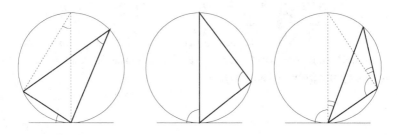

FIGURE A.II

A.3 The Napoleon Theorem

Though the Napoleon theorem is not part of our needed background, the following is an elegant simplification by Kay Hashimoto (a 10th grader at Lakeside School, Seattle) in May 1992, of the proof of Ross Honsberger [*Mathematical Gems*, Mathematical Association of America, Washington, D.C., 1973, pp. 34–36].

THEOREM A.3.1. *On each side of an arbitrary triangle, draw an exterior equilateral triangle. Then the centroids of these three equilateral triangles are the vertices of a fourth equilateral triangle.*

Proof. Given $\triangle ABC$, let X, Y, Z be the centers of the circumcircles of the exterior equilateral triangles on the sides BC, CA, AB, respectively, and O the intersection of the circles Y and Z (other than A). Then, by Corollary A.2.3 in the previous section, we have $\angle AOB = \frac{2\pi}{3} = $

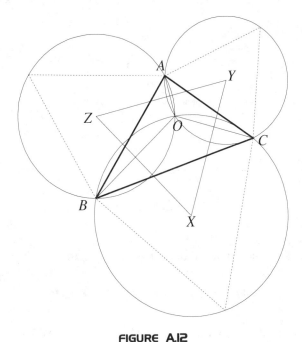

FIGURE A.12

$\angle AOC$. Therefore, $\angle BOC = \frac{2\pi}{3}$. It follows that the circle X also passes through the point O (again, by Corollary A.2.3). We have shown that the three circumcircles meet at the point O.

Now, XY, the line joining the two centers, is perpendicular to the common chord OC. Similarly, XZ is perpendicular to OB. But $\angle BOC = \frac{2\pi}{3}$, and so $\angle X = \frac{\pi}{3}$. Similarly, $\angle Y = \frac{\pi}{3} = \angle Z$, and we are done.   □

A.4 The Apollonius Circle

LEMMA A.4.1. *The interior and the exterior bisectors of an angle at a vertex of a triangle divide the opposite side into the ratio of the lengths of the two remaining sides.*

Proof. Let the interior and exterior angle bisectors at the vertex A intersect the side BC of $\triangle ABC$ at D and at E, respectively. Choose

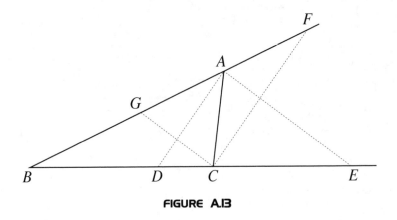

FIGURE A.13

the point F on the extension of the side AB such that $CF \parallel AD$. Then

$$\angle AFC = \angle BAD = \angle DAC = \angle ACF.$$

Therefore, $\triangle ACF$ is an isosceles triangle. It follows that

$$\overline{BD} : \overline{DC} = \overline{BA} : \overline{AF} = \overline{BA} : \overline{AC}.$$

Similarly, choose the point G on AB such that $CG \parallel AE$. Then

$$\angle AGC = \angle FAE = \angle EAC = \angle ACG.$$

Therefore, $\triangle ACG$ is an isosceles triangle. It follows that

$$\overline{BE} : \overline{EC} = \overline{BA} : \overline{AG} = \overline{BA} : \overline{AC}.$$

$\square$

Alternate Proof. We use the same notation as in the above proof. Since D is on the bisector of $\angle ABC$, the perpendiculars from D to AB and AC have the same length (by Lemma A.1.7). Therefore, the ratio of the areas of $\triangle ABD$ and $\triangle ACD$ is $\overline{AB} : \overline{AC}$. On the other hand, these two triangles have common height from the vertex A. Therefore, the ratio of the areas of these two triangles is also equal to $\overline{BD} : \overline{CD}$.

$$\therefore\ \overline{BD} : \overline{CD} = \overline{AB} : \overline{AC}.$$

As for the exterior angle bisector AE at the vertex A, consider $\triangle ABE$ and $\triangle ACE$, and carry out the same argument. $\square$

The converse of the lemma follows from the uniqueness of the point dividing the side of a triangle into a fixed ratio.

COROLLARY A.4.2. *Let D, E be the points on (the extension of) the side BC of $\triangle ABC$ such that*

$$\overline{BD} : \overline{DC} = \overline{AB} : \overline{AC} = \overline{BE} : \overline{EC}.$$

Then AD and AE are the bisectors of the interior and exterior angles at the vertex A.

THEOREM A.4.3 (Apollonius). *Consider a pair of points A, B and a fixed ratio $m : n$. Suppose C and D are the points on the line AB such that*

$$\overline{CA} : \overline{CB} = \overline{DA} : \overline{DB} = m : n.$$

Then a point P is on the circle having CD as its diameter if and only if

$$\overline{PA} : \overline{PB} = m : n.$$

Proof. Suppose P is a point satisfying the condition

$$\overline{PA} : \overline{PB} = \overline{CA} : \overline{CB}\ (= \overline{DA} : \overline{DB}).$$

Then, by Corollary A.4.2, PC, PD are the bisectors of the interior and the exterior angles at the vertex P of $\triangle PAB$. Hence $\angle CPD = \frac{\pi}{2}$, so the point P is on the circle having CD as its diameter.

Conversely, suppose P is an arbitrary point on the circle with CD as its diameter. Choose the points E, F on (the extension of) AP such that $BE \parallel CP$, $BF \parallel DP$. Then

$$\overline{AP} : \overline{PE} = \overline{AC} : \overline{CB} = m : n,$$

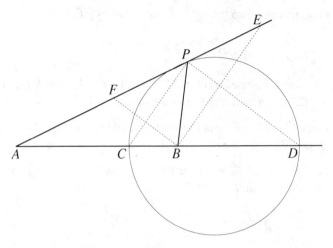

FIGURE A.14

and

$$\overline{AP} : \overline{PF} = \overline{AD} : \overline{DB} = m : n.$$

Therefore, $\overline{PE} = \overline{PF}$. Since $BE \parallel CP$, $BF \parallel DP$, and $\angle CPD = \frac{\pi}{2}$, we have $\angle EBF = \frac{\pi}{2}$. Hence P is the midpoint of the hypotenuse of the right triangle BEF. It follows that $\overline{PB} = \overline{PE}$. Therefore, $\overline{AP} : \overline{PB} = m : n$. $\qquad\qquad\square$

APPENDIX B

New Year Puzzles

The author has been sending New Year puzzles as season's greetings for the past several years. As the purpose is to popularize mathematics, these puzzles are not intended to be hard (except possibly in 1986). Since these puzzles are gaining popularity among the author's friends, we publish them here hoping readers will do the same.

1985

1.
$$0 = (1 - 9 + 8) \times 5 \qquad 1 = 1 - \sqrt{9} + 8 - 5$$
$$2 = 1 + (-\sqrt{9} + 8)/5 \qquad 3 = -1 - 9 + 8 + 5$$
$$4 = 1 \times (-9 + 8) + 5 \qquad 5 = 1 - 9 + 8 + 5$$
$$6 = 1 \times (9 - 8) + 5 \qquad 7 = 1 + 9 - 8 + 5$$
$$8 = ? \qquad 9 = \sqrt{-1 + 9 + 8} + 5$$
$$10 = (1 + 9 - 8) \times 5$$

Can you find a similar expression for 8? (Only additions, subtractions, multiplications, divisions, square roots, and parentheses are permitted. The solution is not unique.)

2.
$$\boxed{}\boxed{}\boxed{}\boxed{}^2 = \boxed{}19\boxed{}\boxed{}85\boxed{}.$$

3. (a) The square of an integer n starts from 1985 :

$$n^2 = 1985 \cdots$$

Find the smallest such positive integer n.

(b) Is there an integer whose square ends with 1985?

1986

Solve the alphametic problem :

$$\begin{array}{r} \text{HAPPY} \\ - \ \text{TIGER} \\ \hline \text{YEAR} \end{array}$$

under the conditions that

1. *TIGER* being the third in the order of 12 animals (rat, ox, tiger, rabbit, dragon, snake, horse, ram, monkey, cock, dog, boar), the number represented by *TIGER* divided by 12 gives a remainder 3;

$$TIGER \equiv 3 \quad (\text{mod } 12); \qquad \text{and}$$

2. as there are 10 possible digits 0, 1, 2, 3, 4, 5, 6, 7, 8, and 9 to fill in the 9 letters that appear in this alphametic problem, there is bound to be one digit missing. However, the missing digit turns out to be the remainder if the number represented by *YEAR* is divided by 12.

1987

Fill in the blanks with digits other than 1, 9, 8, 7 so that the equality becomes valid:

$$\frac{\boxed{}\,1\,\boxed{}\,9\,\boxed{}}{\boxed{}\,\boxed{}\,\boxed{}} = 87$$

1988

1.

$$1988 = 12^2 + 20^2 + 38^2 = 8^2 + 30^2 + 32^2$$

$$= 8^2 + 18^2 + 40^2 = 4^2 + 26^2 + 36^2$$

$$= 4^2 + 6^2 + 44^2 = \boxed{}^2 + \boxed{}^2 + \boxed{}^2 \, ;$$

i.e., find another expression of 1988 as a sum of squares of three positive integers.

2. Show that 1988 cannot be expressed as a sum of squares of two positive integers.

1989

Observe that

$$1989 = (1 + 2 + 3 + 4 + 5)^2 + (3 + 4 + 5 + 6 + 7 + 8 + 9)^2.$$

Find 4 consecutive natural numbers p, q, r, s, and 6 consecutive natural numbers u, v, w, x, y, z, such that

$$1989 = (p + q + r + s)^2 + (u + v + w + x + y + z)^2.$$

1990

Let

$$P_n = 2191^n - 803^n + 608^n - 11^n + 7^n - 2^n.$$

Then

$$P_1 = 1990, \quad P_2 = 4525260 = 1990 \cdot 2274.$$

Prove that P_n is divisible by 1990 for every natural number n.

1991

1. In a magic square, the sum of each row, column and diagonal is the same. For example, Fig 1 is a magic square with the *magic sum* 34. Fill in the blanks in Fig 2 to make it a magic square.

1	8	13	12
14	11	2	7
4	5	16	9
15	10	3	6

Fig 1

	19	91
99		

Fig 2

2. Can an integer with 2 or more digits, and all of whose digits are either 1, 3, 5, 7, or 9 (for example, 1991, 17, 731591375179, 753 are such integers) be the square of an integer?

1992

Choose any five numbers in Fig 1 so that no two of them are in the same row nor the same column, then *add* these five numbers, you will always get 1992. For example,

$$199 + 92 + 177 + 979 + 545 = 1992.$$

19	92	60	665	470
333	406	374	979	784
94	167	135	740	545
199	272	240	845	650
136	209	177	782	587

 Fig 1

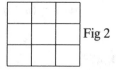 Fig 2

Fill in nine distinct positive integers into Fig 2 such that if you choose any three numbers, no two of them are in the same row, nor the same column, and *multiply* them together, then you will always get 1992. How many essentially different solutions can you find? [Two solutions are considered to be the same if one can be obtained from other by some or all of the following: (a) rotations, (b) reflections, (c) rearrangement of the order of the rows, (d) rearrangement of the order of the columns.]

1993

Let

$$Q_n = 12^n + 43^n + 1950^n + 1981^n.$$

Then

$$Q_1 = 12 + 43 + 1950 + 1981 = 1993 \cdot 2,$$

$$Q_2 = 144 + 1849 + 3802500 + 3924361$$

$$= 7728854 = 1993 \cdot 3878,$$

$$Q_3 = 1728 + 79507 + 7414875000 + 7774159141$$

$$= 15189115376 = 1993 \cdot 7621232.$$

Determine all the positive integers n for which Q_n are divisible by 1993.

1994

We have a sequence of numbers which are reciprocals of the squares of integers 19 through 94:

$$\frac{1}{19^2}, \frac{1}{20^2}, \frac{1}{21^2}, \ldots, \frac{1}{93^2}, \frac{1}{94^2}.$$

Suppose any pair, a and b, of these numbers may be replaced by $a + b - ab$. For example, two numbers $\frac{1}{32^2}$ and $\frac{1}{66^2}$ may be replaced by a single number $\frac{163}{135168}$, because

$$\frac{1}{32^2} + \frac{1}{66^2} - \frac{1}{32^2} \cdot \frac{1}{66^2} = \frac{163}{135168}.$$

Repeat this proceduare until only one number is left. Show that the final number is independent of the way and the order the numbers are paired and replaced. What is the final number?

INDEX